DATA PROFESSIONALS AT WORK

Malathi Mahadevan

Apress®

Data Professionals at Work

Malathi Mahadevan
Raleigh, NC, USA

ISBN-13 (pbk): 978-1-4842-3966-7   ISBN-13 (electronic): 978-1-4842-3967-4
https://doi.org/10.1007/978-1-4842-3967-4

Library of Congress Control Number: 2018960288

Managing Director, Apress Media LLC: Welmoed Spahr
Acquisitions Editor: Jonathan Gennick
Development Editor: Laura Berendson
Coordinating Editor: Rita Fernando

Distributed to the book trade worldwide by Springer Science+Business Media New York, 233 Spring Street, 6th Floor, New York, NY 10013. Phone 1-800-SPRINGER, fax (201) 348-4505, e-mail orders-ny@springer-sbm.com, or visit www.springeronline.com. Apress Media, LLC is a California LLC and the sole member (owner) is Springer Science + Business Media Finance Inc (SSBM Finance Inc). SSBM Finance Inc is a **Delaware** corporation.

For information on translations, please e-mail rights@apress.com, or visit http://www.apress.com/rights-permissions.

Apress titles may be purchased in bulk for academic, corporate, or promotional use. eBook versions and licenses are also available for most titles. For more information, reference our Print and eBook Bulk Sales web page at http://www.apress.com/bulk-sales.

Any source code or other supplementary material referenced by the author in this book is available to readers on GitHub via the book's product page, located at www.apress.com/9781484239667. For more detailed information, please visit http://www.apress.com/source-code.

Printed on acid-free paper

To my soul sisters, Shanthi and Chitra

Contents

About the Author

Malathi Mahadevan is a senior database engineer with more than 20 years of experience working with data, primarily using Microsoft SQL Server and related technologies. She has worked in many industries, such as health care, finance, and consulting, to name a few. Mala has volunteered with the SQL Server community by arranging free training and seminars for the past 15 years. She is a recipient of the PASSion award for outstanding volunteer from the Professional Association of SQL Server (PASS). She blogs frequently at http://curiousaboutdata.com and is active on Twitter as @sqlmal. Mala is also a featured blogger on SQLServerCentral.com and has written several articles for the site.

Foreword

We are living in an era in which data professionals are experiencing a Cambrian-like explosion in their diversity, importance, and value in the workplace. The Cambrian explosion was an event approximately 541 million years ago in the Cambrian period when most major animal phyla appeared in the fossil record. Following that, a multitude of important, world-changing phenomena have happened, such as the establishment of almost all major animal phyla, diversification. Speciation accelerated at lightning speed, and the earth soon bristled with life in every nook, niche, and cranny of the planet. The old world of simple, single-celled organisms and the occasional colony-creature gave way to environments full of complex, large, and colorful life forms not very unlike our world today.

In the same way, where there were once a mere handful of professions in an industry known as "data processing," we are evolving to a world in which even the single, distinct job of yesteryear known as "programmer" has splintered and speciated into dozens of distinct careers. Just as with the example of early life forms, the behaviors and patterns used by early programmers are now divided into a wide range of assorted and distinct groups—from front-end Java programmers, to iOS game developers, to agile managers and scrum masters, to Azure stateless programmers, to back-end SQL Server developers. All distinct. All relying on different skills to succeed in their environments. But that is just one example of many careers in a world that has transitioned from the old days of data processing; an antiquated world focused on computerizing and speeding only the work of scientists and accountants, into the new world of information technology, in which we are embarking on a mission of complete digital transformation of all forms of human activity.

So where does that leave the data professional?

This book will answer that question for you on two main levels.

The first way in which this book functions is to offer a distillation of the institutional knowledge of our industry's best minds. To date, institutional knowledge is still a weakness in our industry. Frequently, our teams are small. There are relatively few members on data-centric teams—say a team of database administrators (DBAs), even within large corporations and organizations. As a result, there is little time to document and accumulate the shared experiences and best practices of our profession into a codified resource for future team members.

Teammates come and go, taking their knowledge and experiences with them. What they knew that made the organization run well and thrive left with them. And the remaining staff must now contend with huge gaps in their knowledge, muddle through challenges for which they have no previous experience, and encounter new and unforeseen circumstances that put them under duress and enormous stress. It's not pretty when a key teammate leaves and the remaining teammates don't know how to do what s/he did for the organization.

On the other hand, institutional knowledge is a kind of shared knowledge that bolsters and improves the team's performance and the wider organization as a whole. It can take many forms, from frameworks, to specific step-by-step process, to finding the best tools for the job, to tips on how to navigate specific relationships within the organization that facilitate best results when implementing a given project. Institutional knowledge, when properly shared across a team, lessens learning curves, helps you avoid pitfalls, and enables you to exploit the successes of others so that you can be both more efficient and more effective. Institutional knowledge can also provide a way forward in uncertain times; essentially, they can provide a process to follow. When the leader is out of the picture, you now have an answer to the question "What do I do now?" offering you not only a set of steps to follow, but an expectation of what priorities to follow.

The second way in which this book functions is to share the stories—complete with protagonists and antagonists—of the human journey to achieve mastery in a profession in which there are still many trails to be blazed. These stories can help the newcomer decide if this is a professional path you would like to follow. For those already on this path, these stories will help you assess your own experiences and, where appropriate, will enable you to incorporate the victories (and mistakes) of the interviewees of this book into your own corpus. That process of incorporation will, in turn, elevate your own mastery and the trajectory of your career.

When people decide to pursue a career as an aircraft pilot, they are embarking on a professional career in which there is a checklist for literally *every* conceivable problem scenario. There are many professions like that in today's world. That's not the world of a data professional. Data professionals are much more like the creative classes of writers, artists, composers, and architects. There are many shades of gray in this profession. What works well for one DBA might not work for another. Combining one group of people to deliver a well-defined project may produce an entirely different result compared to a different group of people. These stories help you to see the human element and the art-like beauty that is inherent in this very creative class. You will see, in these pages, our writers struggle like characters in a drama out of literature, not only with their environment or with an adversary, but also with themselves. Overwork, burnout, imbalances in family life and work life—all of these are relevant to the profession. You will see, in these pages, where our writers have moments of epiphany and then, by incorporating their revelations, achieved breakthrough performance in their career. You will see, in these pages, times of community joined and enjoyed, and of community missed. You will see, in these pages, the influence of many of the luminaries of our profession who, whether physically or virtually, helped train up these writers as if they were in an apprentices-master relation of previous eras. The human journey, then, is the second major offering of this book.

I encourage you to read on and learn from these stories, as you would from any book about technology. But I also exhort you to read them as human stories, of people who strove and struggled and, eventually, succeeded. Learn from them and make their forward progress part of your own.

Kevin Edward Kline

Principal Program Manager, SentryOne

Microsoft Most Valuable Professional (MVP) since 2003; PASS Founder Board Member, President (2004-2008), Lifetime Achievement Award (2010)

Acknowledgments

My thanks should go, in no small measure—to the SQL Server community, fondly called SQL Family. My thanks to all who are part of it but some are special because they have been part of my journey and helped me in special caring ways. Their support and wise counsel has helped me stick with community work. Those special people happen to be Kevin Kline, Karla Landrum, Joseph Sack—and the PASS organization in general. Thank you for being there for me.

A sincere thank you to fellow geeks and beloved friends who are no longer around—Tom Roush, Robert Davis and Dave Ingram. Your lives and your love for what you did will continue to inspire us all.

I must thank Kathi Kellenberger, for encouraging and inspiring me on my writing goals—This book would not have happened without her.

Thank you for the kind and considerate people at Apress who went through so many iterations of this with patience and diligence—Rita Fernando, Kim Burton-Weisman, and Jonathan Gennick, it was a privilege working with you.

Last but not the least, the people in my family: my second mother and beloved aunt, Rajam—your persistence, creativity, and hearty laughter are my greatest inspiration. And the two people I love the most, my soul sisters, Shanthi and Chitra. You mean everything to me. Thank you for your love and support.

Introduction

I started my career as a database architect maintaining entity relationship diagrams for a software company many years ago. I maintained such diagrams for all of their customers. I made changes as their systems changed, printed as necessary, and created new ones when new databases came up. I sat through meetings, discussing database schema, normalization, and so on. Back in those days, every software development team had an architect like me. A few years later, things changed. Developers started designing their own schemas. Fancy graphical user interfaces began to maintain them. Architects were deemed a luxury and not a necessity. Many people like me turned to programming to sustain our jobs.

Then, more database administration jobs started to come up, and I took one of them. There was no roadmap to a career in data back then; everyone just figured it out as they went along. To some extent, it is like that now too. But now, there are a lot more people who have walked this path. We have a better idea now of how things can change, and which skills we may need to brace ourselves for change.

We are also aware that there are many possible tracks to follow in the vast area of working with data. Each track has possibilities of a different kind - for example, a business intelligence track may lead to a career in data science, and a database administration track may lead you to become a cloud architect. None of these are guaranteed paths, of course; life is a complicated journey and anything may lead anywhere. But, a track is a good guideline to follow, because there are people who have followed them and have been successful at what they do.

I've had the good fortune of knowing many talented, successful people as a result of being part of the SQL Server community. A few of these people are in this book. The people I've interviewed are rock stars at what they do. They are also very giving and generous, as I have known personally from many years of being in the community. Their stories will help anyone looking to become a DBA, BI professional, data architect, or database developer/engineer by offering tips and suggestions on how to be better at what they do. If you are reading this, my guess is that you are a person with a career in data or looking to become one.

There is no set way to read a book like this. You may know some of these people from their blogs or Twitter tags, and then chose to read their stories. Consider what kind of "track" you are on, and pick the right chapter to read. The following are some guidelines for this.

Are you a DBA or looking to be one?

The DBAs interviewed in this book are Andy Mallon (Chapter 12), Drew Furguiele (Chapter 24), Ken Fisher (Chapter 3), Marlon Ribunal (Chapter 26), and Tracy Boggiano (Chapter 21). Read their stories to understand how they got to be great at what they do.

Are you an experienced DBA and looking to move on to consulting?

Jason Brimhall (Chapter 10), Jes Borland (Chapter 5), Joe Fleming (Chapter 27), John Morehouse (Chapter 16), and Mindy Curnutt (Chapter 1) are DBAs who have graduated to become successful consultants.

Are you a person with considerable experience under your belt and looking to be an architect, or are you an architect looking to be a better architect?

Argenis Fernandez (Chapter 19), Jimmy May (Chapter 24), and Dave Walden (Chapter 22) are among the most respected data architects who have shared their stories.

Are you already dabbling in business intelligence or looking to learn more about BIML, data visualization, and so on?

There are no better people than Andy Leonard (Chapter 4), Julie Smith (Chapter 2), Jonathan Stewart (Chapter 14), Kirsten Benzel (Chapter 20), and Tim Costello (Chapter 11) to understand this path.

Are you leaning toward data science and analytics as a possible career choice?

Learn from Kevin Feasel (Chapter 6), Steph Locke (Chapter 13), Ginger Grant (Chapter 7), and Matt Gordon (Chapter 23) on how they got there!

Are you looking at being a product manager after having gained some mastery over a product, or perhaps certain features of a product?

Learn from some of the best product managers—Joseph Sack (Chapter 15), Vicky Harp (Chapter 8), and John Martin (Chapter 15)!

Lastly, are you looking to do something totally different, like teach or become a tech editor?

Kendra Little (Chapter 9) and Kathi Kellenberger (Chapter 18) can share their stories in this regard.

This is just a very rough guideline. Keep in mind that all of these people have had a mix of skills over the course of their journeys, as we all do. It is worth reading all of their stories to understand what led them to where they are now. I hope you learn and grow from them as I did. Happy reading!

Mindy Curnutt

CEO, Mindy Curnutt and Associates Consulting

Mindy Curnutt has been working with data since the early 1990s and with SQL Server specifically since 1997. She is CEO of Mindy Curnutt & Associates Consulting. She is Vice President of Strategic Partnerships for the nonprofit startup Girls + Data (www.girlsanddata.org), where she is responsible for fundraising, social media promotion, and strategically advancing the company's mission to encourage young women to pursue careers in data analytics. Mindy also owns Kirbson Direct, LLC, an umbrella over two other companies Stampola (www.stampola.com) and BBQ Bud (www.bbqbud.com). Stampola manufactures rubber stamps and silicone molds. BBQ Bud designs, manufactures, and imports novelty barbeque accessories.

Mindy holds a BA in business economics from the University of California, Santa Barbara. She is president of the board for the North Texas SQL Server User Group. She served five years on the PASS Summit Program Committee—three of them as a manager. Mindy has been awarded the Microsoft Data Platform MVP designation four times. She has been PASS Volunteer of the Month twice. Mindy is a co-author of three books: Let Her Finish: Voices from the Data Platform (CreateSpace, 2017), SQL Server 2017 Administration Inside Out (Microsoft Press, 2018), and Microsoft SQL Server 2012 Step by Step (Microsoft Press, 2014). She provided insight as a subject matter expert for the SQL Server 2012 and 2014 certification exams.

© Malathi Mahadevan 2018
M. Mahadevan, *Data Professionals at Work*,
https://doi.org/10.1007/978-1-4842-3967-4_1

An avid hobbyist, Mindy enjoys crocheting animals and figures (amigurumi), making soap from oil and lye for Riley Roo Soap Company (www.rrsoap.com), baking things for her two children, Riley and Kimball, and playing guitar and performing at the Sons of Hermann Hall in Dallas, Texas, with her husband, Chris. You can find Mindy online at www.mindycurnutt.com or on Twitter as @sqlgirl.

Mala Mahadevan: Describe your journey into the data profession.

Mindy Curnutt: I started my college experience at UC Santa Barbara as a nuclear engineering major. That lasted about a semester. I had imposter syndrome. There were very few females in the classes, and I was also one of very few Caucasians in the classes. Granted, the material was difficult, but I was also intimidated by being different from everyone else.

To make things worse, I did not know how to study. See, I had breezed through high school with very little effort required. When I got to college, it became very apparent that I didn't actually know how to study! I ended up getting some pretty low grades that first semester, and it really freaked me out. This was in addition to the whole impostor syndrome problem. Other students in my dorm hall were taking fun classes that I did not get to take because I was in the engineering school. I was interested in archaeology and old music history classes, and so on. I decided to drop out of the engineering school and enroll in the art school so that I could take these more interesting classes and be more comfortable.

Over the next few years, I ended up changing my major multiple times. Then it came time to graduate. I was pressured by my grandmother because she felt I wasn't going to be graduating with a degree that was marketable. She kept asking me, "What are you going to do for a living?" At the last minute, I changed my major *once again* and ended up getting a degree in economics. It was actually very interesting—much to my surprise! Statistics was one of the classes I had to take, and I ended up receiving the highest grade of my entire college career *and* out of the entire class in that subject. Before taking that class, I didn't even know what statistics were. It was really strange. Right there was probably the first clue that I was going to end up in the data field.

So, I graduated with a degree in economics, but right about then [1991], the economy wasn't very good and I could not get a job for the life of me. I sent out hundreds of résumés. I did not want to live at home with my parents. I ended up getting a job in the accounting department of a car dealership. From there, I ended up moving back to Napa Valley, where I was from. I got a job working for an economist, helping him create supply-and-demand reports for the wine grape industry. I did a lot of data entry, charts, and graphs using Lotus 1-2-3 in DOS. I don't even remember what the database platform was. It was some DOS-based, DB2 type of thing. I also did a lot of mean, median, and statistical variation. We were making charts using a dot matrix printer that had these different colored pens I had to insert into clamps. That was my first work with databases and computer hardware.

Then I traveled with a music band to Alaska and found my way to Seattle. That's a long story, not for here. Seattle is where I ended up getting into SQL Server. I worked for a furniture manufacturing company, and I developed a just-in-time manufacturing system for them using Access 2.0. They bought another system to manage things, which ran on SQL server 6.5. I migrated the database that I had written to the new SQL Server platform. I was sent to training on SQL Server, and it just turned into a whole career. It was not planned, but in hindsight, I ended up doing the database stuff because I was just naturally drawn to it and good at it. I like organizing things. I like numbers. I like making sense of data. I like puzzles. A match made in heaven.

Mala: Describe a few things that you wish you knew when you started your career that you do know now and would recommend to people.

Mindy: I did not reach out to the SQL community and get involved in the SQL family early enough. I really only did that in the last ten years. The community has grown a lot in the last decade, but there are a lot of people that you look to—especially the old-timers. People like Kevin Kline, Kimberly Tripp, Kalen Delaney, Allen White, and Aaron Bertrand have been around for twenty years or more. I haven't. They've known each other a long time. I didn't religiously go to the conferences. I didn't make sure that people knew me. That association with other people who are peers who do the same sort of work that you do is super important to building your brand.

Mala: Yeah, I was like that too. I am one of the old-timers, but nobody even knew I existed for a very long time. I just did not know how to do that.

Mindy: Yeah, exactly right. There is a huge career opportunity out there. It is a good thing for you to get to know other people who work in the same field that you do. In the SQL community, there are some really good people who can help you get to where you need to be. I've heard that some of the other communities are not the same way. This is a very special community—the #SQLFamily. That is something that I wish I could go back to the younger me and say, "Get involved."

Mala: Describe a day in your professional life.

Mindy: I now work for myself. I've been self-employed successfully now for six months. Today it's New Year's Eve. I am actually comparing side-by-side traces of the exact same database with the exact same replay on different versions of SQL. I forget what day of the week it is now. I work when I have some work to do. And when I don't have work, I don't work. On a typical day, I get up and make breakfast for the kids, see my son off to school. My husband takes my son to school or my son rides his bike. My daughter is in high school, so she goes to school an hour later. I sit down in my pajamas—my hair still completely a disaster—with my coffee. Then I start checking e-mail and catching up to see what happened. I get started on projects. Around lunchtime, I'll get a shower, wash my hair, take the dog for a walk. The kids come home about 3:30, and then they've got to be carted to a lesson of some sort. Then at night, I usually do some more work—sometimes with a glass of wine! It is very different from when I was working for a corporation. I don't have as many meetings. I've actually been able to set time aside and do some online classes too. In my life, I've gotten to the point where I have networked enough and have created demand, which allows me to be self-employed. It's been pretty nice.

Mala: So glad to hear that. I hope to get there someday. What are some trends in this line of work that you're watching for?

Mindy: Thirty to thirty-five percent of what I do is performance tuning. SQL Server 2017, with its self-tuning abilities, is something I look forward to. I like that it can track parameter sniffing on its own. And that it can automatically adjust using execution plans tailored to specific parameters and performance over time. The building of that kind of artificial intelligence into the database engine is awesome.

If you are using Azure SQL Database, you don't need to worry about backups. You don't need to worry about setting up all the high availability and disaster recovery. It is all built in, which eliminates some of the need for what I do. I am working with the real estate industry. I worked with the trucking industry for the past several years. A lot of those companies still aren't on SQL Server 2016. There aren't even in 2012. Even though a new version of the on-prem SQL engine seems to be coming out every single year, a lot of people are on older versions and are not moving forward as quickly.

Still, I do need to start focusing on modernizing and upgrading my skill set to where the world is headed. I'm interested in expanding my skill set in visual analytics. I have an accounting background and own two other businesses besides this. I understand what business owners want to see. I understand budgets. I understand that revenues drive things. I get the connection between the data that we're collecting and the data that we aren't collecting. What we should be collecting isn't there right now. And I know what we could be doing with that information if it were available. Like what kind of dashboards we can provide to give better visibility to the business owners and the business drivers. That's where I want to go with this technical knowledge. I'm not just somebody who's got data skills and makes cool Power BI stuff. I can come in and be a full consultant that not only understands your data platform needs and what's going on with your databases, but how can you use that data. Also, I provide information on what other data you can collect that you are not collecting, and what you can do with that data as a company. I'm going to be in my fifties soon, so I think that that's a good place for me to go because it's very advisory.

Mala: Very interesting. Getting to fifty changes your perspective on a lot of things. It did for me. I used to value job security most. I now am looking more and more at what I want to enjoy doing. I realized in a very real way that my time is limited, and if I want to do what I enjoy, it is now.

Describe a few things that a database professional should follow as best practices.

Mindy: The number-one best practice is knowing how to restore your system. I can't even tell you how many people can't do that one thing. And they aren't even aware of the danger that presents or the severity of the situation. It may sound like beating a dead horse, but people lose their data *all the time.* I was at a company the other day and I asked the guy, "Do you know where your backup is?" He said, "Well, we have somebody who takes care of that." I asked, "Who?" And he said, "We have an offsite IT company, and they've assured me that that's all taken care of." I asked him, "Are they keeping copies offsite?" He said, "I don't really know." Incredible. You've got to know.

Mala: I worked for a private hosting company once. They had horror stories of people like that. They don't know that they need to know—until their stuff is actually completely lost.

Mindy: Yeah. You have to start asking the questions. How long could you live without this data? Would your company be out of business, or you could you hold your breath for a week and survive with no access to your systems? How bad of a hit would it be if it took two weeks?

It's not just database corruption. It can be fire or theft. Or even somebody going outside and severing all the fiber optic lines into the building accidentally with a backhoe. I've seen that happen. Yeah. "Well, actually our database is fine but we have absolutely no Internet." That can be a problem.

I even have a story of a guy who got really mad, came in and kicked the server room door in, poured gasoline on servers, and lit them. That is unlikely, but you could have a cable cut by an outsider, or some sort of a major grid failure, or something where you don't have electricity. You have a generator, but it can only run for a day. To me, that's the number-one best practice: be recoverable.

The second is to be organized. What I find for myself—and I still struggle with this—is that sometimes I'm moving so fast that I don't document exactly everything that I'm doing. And I find that if I force myself to do stuff with a script instead of using the GUI [graphical user interface], then I have it documented. The script self-documents exactly what took place. No question of did I check that box or did I not check that box in the script. Also, if you if you're doing it via script, you can reuse it. That saves you time the next time.

Have an organized library. I have a laptop, four different e-mail accounts, and some VMs [virtual machines] out on Azure. I've also got external drives all over the place—in this backpack or that purse. If I'm looking for just the right script, I better know where I put it. Having a system and being organized is the way to go.

Another best and worst practice–especially as you get higher up in your career–is to learn to delegate. I have been a lot better at that in the last three or four years. At a certain point in your career, if you have people that work for you, you recognize their strengths. If they have strengths in areas that you know you're weak or not as strong, then take advantage of their strengths! Lean on them, right? I mean, if they are really good at scripting stuff, or they're really good at error handling, or they're really good at just being consistent about stuff—and I'm not, then find ways to use their strengths to better the whole team. You don't have to do everything yourself.

Mala: Describe your experiences with cloud adoption.

Mindy: I'm starting—more and more—to consolidate my stuff out into the cloud. Five years ago, I would have all my stuff on my laptop, and then I had my laptop backed up to the cloud using Mozy or Carbonite, or something like that. But now I'm trying to be really good about putting my stuff out in the cloud. If you are a consultant, you want to have templates for things— contracts, SOWs [statement of work], scripts, assessments, etc. Things need to be consistent, locatable, and branded. You need customer folders and documentation. What rate did you quote? What timeframe did you say? And being organized is so valuable.

I'll admit that one of my worst habits is shooting from the hip. So, it's a battle, and I have to be very pragmatic. Sometimes I thank myself, and sometimes I really disappoint myself when I can't find things—when I should have just slowed down and been more methodical. I've started using the cloud as a universal place to organize all my things, so I can find them no matter where I am or what device I'm using. The cloud has enabled me to be more organized and more consistent with my internal system of "things" I use in my work.

I think you were really asking about cloud adoption in the sense of SQL on-prem versus Azure data platform offerings, but the cloud storage for all the documents, files, and systems that I've amassed over the years and that I use in my career were the first things that came to mind when you asked that question.

I have seen a lot of enthusiasm for people wanting to "go to the cloud." When I'm talking about the cloud, I'm talking specifically about the data platform being in the cloud. I think people are still confused. I think they want to do it [move to the cloud] because they think they *should* do it, but they can't tell you *why*. I don't think everyone wants to do it for the right reasons. Historically, an application's database platform might be Oracle, or SQL Server, or DB2, or whatever—and be on-prem. With these newer IoT data streams and data connecting through many APIs, having one database platform as the "be all and end all" is no longer realistic.

You may have two, three, four, or five different types of data platforms supporting your architecture. You may have some stuff that is relational and normalized. You may have some stuff that is just static logs, and other stuff that is has graph-like relationships. You may have some stuff that you're saving "just in case" but currently serves no purpose. It is a combination of data sources, types, variability, longevity, stasis, and velocity. That's a new world for many of us.

Many people want to move to the cloud because they think that's the future. They are right about that, but they're prematurely trying to push everything up there just so that they can say that their company is "in the cloud." I'm seeing a lot of that. I hope there is not going to being a backlash. In many cases, the platform is decided upon before they even know what the problem is.

Mala: What are your thoughts on agile and database administration?

Mindy: Have tools. You have to have the database in some sort of source code control with a toolset, or you'll make yourself crazy with agile. My experience working for a software vendor is that you have to know what version of the database goes with the build. And if you can't do that, then I don't know how you're going to manage working with agile.

Mala: What are some of your favorite tools and techniques?

Mindy: SentryOne Plan Explorer is my favorite tool, hands down. I could be an evangelist for that tool all day long.

One of the techniques that I actually teach—but I've never seen anyone else demonstrate—is to do a wide-open extended event capture or trace, and then write a script that detects time between the completion of an event and the start of the next event. The *missing time* shows the time that the user or process is experiencing that is not time spent on SQL Server. You won't find that figure calculated anywhere in wait stats or in any of the many monitoring performance tools out there. I call it "reading between the lines."

Mala: What are your favorite books, blogs, and other means of learning?

Mindy: I read a lot of books at the same time. When I'm going to go to bed, I'll just read a half of a chapter or a chapter of a book. I don't read fiction. *The 4-Hour Workweek* by Tim Ferris [Harmony, 2009] is an awesome book. You can read this in tiny chunks. The chapters are all independent. It talks about how to delegate things, and that you don't have to do everything yourself to be successful. It suggests batching your tasks, like read your e-mail twice a day. Don't read your e-mail all day long, every second as it comes in. That is super inefficient and makes you crazy! All that context switching wastes an enormous amount of time. The suggestions in that book are meant to help you take your forty-hour workweek down to half that or even less, but still get the same amount done.

Switch by Chip Heath and Dan Heath [Crown Business, 2010] is another self-help book. It is about how to change things when changing things is hard.

A good book on public speaking is *Steal the Show* by Michael Port [Mariner Books, 2016]. It talks about how to do good speeches and job interviews.

I have also been meaning to read "PowerShell in a month of lunches". It's been on my desk for six months now. I've had a lot of people telling me it's great.

Admittedly, I'm not a big blogger, and I don't follow too many blogs.

Mala: What do you recommend by way of stress management—handling long hours on the job and staying motivated?

Mindy: Exercise is a big thing for me. I try to take my dog for a walk every day. I use a Fitbit to track my steps. There is a really large SQL community of people that use the Fitbit. You can do a week challenge with them. It keeps pinging you and showing you that now you're in seventh place or something, and you have to get up and start moving or you're going to be the loser! We can get fiercely competitive with each other on who won the week or weekend challenge.

I need quiet time where I'm not plugged into any electronic devices too. When I'm walking, I don't even listen to music. I just walk. I need that sort of downtime where I'm just listening to myself breathe, and hanging with my dog. And sort of focusing on what's happening inside, because there's so much digitally every day that just keeps us just spinning.

Mala: What are typical questions that you ask in an interview? What do you expect of a new DBA?

Mindy: If somebody is claiming that they're a DBA, and they have all this experience, the résumé looks great and all that. The first question that I would ask is, "In the latest version of Azure DB, what's the maximum number of clustered indexes that you can have per table?"

Mala: Ha!

Mindy: Yes! So if they don't immediately do what you just did, then they're not really a DBA. We can just stop the interview right now.

If we get past that, I basically like to give somebody two or three problems. They have an hour. And I'm not going to be looking over their shoulder. Here's a computer and here's Google. I want to see what solution you come up with. And maybe there's no absolute right answer. Here is a query. Can you make it run faster? How do they go about doing that? Did they figure out that it only runs slowly when you give it certain parameters first and then the other parameters second? I basically want to see that somebody can think on their feet and have tenacity, common sense, decent skills, and can problem solve.

If you are managing a team, it is also really important to make sure that you're adding somebody who is going to have a synergy with the team. If you are a DBA and you are called in on a night call with two people, you have to trust and like them. Right? That they're not a jerk when they are tired or whatever. Personality is a part of it.

Mala: Oh, absolutely. Yeah.

I think that you already answered the last question I had planned. Is there anything more you'd like to say on community, and why you recommend being involved with the community?

Mindy: Go out and get speaking experience. If you live in a large metropolitan area, there are a dot-net user groups. There are all sorts of developer user groups that would probably love to have you come and talk about indexes and how they work. I spoke with three different user groups in Dallas this year that I didn't even know existed beforehand.

Mala: Can you share a funny story from your life?

Mindy: I was fifteen years old, so it was like 1984 I guess. I lived in Calistoga, which is the most northern town in Napa Valley. Robin Williams had a home somewhere mid-valley. My friend Laurie and I were headed home from an event in town. She had a Ford Galaxy convertible that her grandma had given her. We had the top down—California weather, always nice. As we drove by a motel on the outskirts of town, we heard a lot of ruckus and laughter echoing into the night. It was odd since that motel was normally really quiet and dark. Laurie pulled over, and we parked and walked around the back of the motel to see what all the noise was about. There was Robin Williams standing on a picnic table giving an impromptu concert to about twenty people. We sat down on the lawn and enjoyed about fifteen minutes of him just ripping joke after joke after joke, like lightning.

Afterward, Laurie and I went up to meet him and get his signature. I didn't have any paper on me except for my provisional driver's license. I wasn't sixteen yet. I held that up for him to sign on. He saw my name was Melinda and asked me point blank, "Do you go as Mindy?" I said yes. Next thing you know, he's gone crazy and grabs me in a headlock and starts walking around announcing, "Hey? It's Mork and Mindy! Nanu, nanu!" And doing all this *Mork & Mindy* stuff, acting all "Morkish." Everyone was laughing. During that time, anytime someone heard that my name was Mindy, I'd get some kind of *Mork & Mindy* comment. So this was really over the top since it was Robin Williams himself doing it to me.

Mala: Wow, what a fascinating story! Thanks, Mindy.

Key Takeaways

- Be involved in the community. Go out and get speaking experience. Find user groups in your area to go to, and volunteer to share what you know.

- Be recoverable with your data—anything can happen.

- Define the problem before deciding on the platform. Don't find out the hard way that you picked the wrong platform.

- Be organized. Script what you do so that you know what you did, and save it for future use and reference.

- Have tenacity. Be trustworthy.

Favorite tools: SentryOne's Plan Explorer, SQL Server Extended Events

Recommended books: *Switch* by Chip Heath and Dan Heath, *Learn Windows PowerShell in a Month of Lunches* by Donald Jones and Jeffrey Hicks, *Steal the Show* by Michael Port, *The 4-Hour Workweek* by Tim Ferris

Recommended training/conferences: SQLSaturday, PASS Summit

Julie Smith

Business Intelligence Consultant, Innovative Architects Inc.

Julie Smith has been moving and bending data for 15 years using varied tools such as Microsoft Access, MySQL, SQL Server, Azure Data Factory, Azure SQL Database, and Power BI. She has built and designed increasingly complex BI (business intelligence) solutions for organizations across multiple industries in her role as a business intelligence consultant at Innovative Architects in Atlanta, Georgia.

Julie earned a BA in theater and speech from the University of South Carolina. She worked as a costume technician and an actor for several theaters in Atlanta before discovering a love and talent for data. Julie has been a Microsoft Data Platform MVP since 2013. She is a frequent speaker at user groups and technical conferences such as PASS Summit. She blogs at datachix.com and microblogs as @juliechix on Twitter. She contributed several chapters of Microsoft SQL Server 2012 Bible (Wiley, 2012).

Julie lives in Atlanta with her two sons, Edmund and Gareth Wright. She is also building a tiny house in North Carolina with the love of her life, Jim Christopher.

© Malathi Mahadevan 2018
M. Mahadevan, *Data Professionals at Work*,
https://doi.org/10.1007/978-1-4842-3967-4_2

Chapter 2 | Julie Smith: *Business Intelligence Consultant, Innovative Architects Inc.*

Mala Mahadevan: Describe your journey into the data profession.

Julie Smith: I graduated from college with a degree in theater, and I did theater for five years, mostly costuming. Then I realized that I wasn't making enough money to support the children I wanted to have. I'm from Erie, Pennsylvania, originally, and I had watched my dad go from being a welder in a shop who took his love of personal computers—he had a TRS-80—over about three or four years to turn himself into a programmer. So, I had that in my family background, and I knew it could be done.

As I was doing theater and making costumes, I got a job as an administrative assistant. I was the only person at my job who liked using Crystal Reports. They used Crystal Reports for something, and I got into the guts of Crystal Reports and said, "Hmm, this is something that I could do." So, after doing that for a little while, I declared, "I want to make a change. I want to do data." I really didn't even know what all it meant, but I just said, "I want to do data."

This was back when you still looked for jobs in a paper newspaper. I saw an ad for instructor at New Horizons Computer Learning Centers. I thought to myself, "I can teach and this would be a great way to learn what I need to know to make this transition." They liked the fact that I had a theater background because they thought it would make me a good teacher. I got the job. They started everyone out on Excel and Word—the productivity stuff. I was a little bossy-pants, and I said, "No, I'm only going to do Access. At the time, it was a desktop database." Since the Access class was regarded as difficult, there were fewer of the desktop instructors who would teach it, so I got my way. I couldn't teach SQL Server at New Horizons. So, I said, "Well, Access is the closest thing to data," so I insisted on teaching the Access classes. I taught them for a little while, and then I used what I had learned to get a job as a reports analyst at a law firm.

Over the next several years, I paid for my own training. I went to New Horizons as a student and took the SQL Server training, so I did this hybrid thing. After four years at the law firm doing Access, a man named Mark Fugel took a chance on me, and I got my first SQL Server development job. That was in 2006, and from there, I went from junior to mid-level and took on the ETL job at that place. That's how it all started. I am so grateful to him for giving me that chance.

Mala: Wow. Does your theater background still help you in teaching?

Julie: I don't think that the theater background helps me in presenting. I think that the reason that I love doing theater is the same reason that I love presenting. So, it's just kind of like, "Well, take Julie and put her in a theater background, and she's going to enjoy acting. If you put her in a data background, she's going to enjoy giving presentations and trying to make people laugh." I do get a kick out of making people giggle.

Mala: Describe a few things you wish you knew when you started your career that you know now and that you would recommend to people who are newbies.

Julie: Because I came in from a different background, I was really insecure about my knowledge. I see that with people because I've mentored a few people who are trying to make a transition from admin or reports analyst into more technical roles. I think that we all let the fact that we don't have technical knowledge haunt us. I would urge you to rest easy on the fact that the most important thing is marrying the technical knowledge with understanding business needs and understanding what's going on and really solving the problem. I think that that has served me well, even though I didn't consciously know that that was what I was doing. I've always been a person who can go in and say, "Well, what's going on here?"

I had a boss once who said, "In order to be able to automate something, you have to be able to do it manually first." I've been in situations where people are like, "We're going to automate X process," and it's like, "Well hold on, what are you doing?" There's not even an understanding of what you're actually trying to do and little understanding of how to do it manually. You have to do it manually and then figure out how to automate it.

I think people who are senior understand that and will maybe even unconsciously navigate that without knowing that's what they're doing. But people will say, "Oh, should I just go get this certification?" I'm like, "Well, you can get that certification. it certainly won't hurt you. It's great to have that knowledge. But don't lean on it or don't think that you're not ready to be effective at work just because you don't have it."

Mala: That is true. So, do you recommend that people become certified? What is your take on certification, generally?

Julie: I haven't had a certification in a while, and that's probably not a good thing. I think, right now, the technology is changing so fast, at least in the Microsoft stack, that I don't know how the certifications can keep up. Again, it won't hurt you, but there's a chance that you're being tested on stuff that's going to be out of date by the time you take the test.

Mala: What's a typical day in your life as a professional?

Julie: I have different types of days. Some days I am head down in development for my own projects. Some days I have client meetings for presentations or just check-ins. I do a lot of hands-on development still. I also run a team for my client, and I am a team lead at Innovative Architects. I get pinged a lot by other projects at IA for architectural consults. I balance my day with answering, guiding, and leading the decisions on the technology for my client and also for other projects at Innovative Architects. I do something I would call "informal agile," which is basically just gathering requirements for a small project and working on it, and then getting it in front of the client for feedback as soon as possible. I have feedback sessions with the client. "Does this work? Is this what you need? Yes, no?" Then we go from there and just plan the projects and do that in an informal agile way.

Mala: What are some interesting trends that you are following in technology?

Julie: We are moving to cloud technology, and that is snowballing. Most of my career has been doing SQL Server Integration Services. The thing that's exciting to me is the emergence of artificial intelligence and what it's doing to that field. There are products that are using big data technology and Spark. They're able to take what used to be totally manual processes and requirements gathering, and do data cleansing. So, there are products out there—the Trifacta and Paxata, and now Azure Databricks—that are based on these two technologies. They really help because they do data cleansing automatically, just based on algorithms that are internal. It doesn't necessarily take a person to say every iteration, every version of a word.

You're in Kentucky, right?

Mala: Yeah.

Julie: Yeah, so Kentucky with a space after it, Kentucky with a period, KY, all of those variations. Or maybe even seeing that the problem isn't that it's just the word Kentucky—it's a bunch of words. It would take all that and just cleanse it automatically with almost no human intervention.

Mala: That's awesome. Are you using any of those things now?

Julie: I'm not using them in my work currently. I have evaluated them, and I always go back to them because it's just so exciting. But I joke frequently that I was happy to have my role change. I said I was happy to be out of an ETL job because the yucky parts of the job became automated. You know, because extract-transform-load data transformation is a lot of drudgery in many ways, no matter what. It's kind of the riskiest part of the business intelligence stack. Things often fail in the extract-transform-load process of a data warehouse project. So, I'm happy to have help. I'm happy to move on to cubes and other parts of the stack.

Mala: So that's interesting. I was just hearing on a podcast yesterday that SSAS [SQL Server Analysis Services] has more of a future now compared to SSIS [SQL Server Integration Services].

Julie: Interesting.

Mala: Yeah. I don't know the complete details, but I thought what you said seemed to fit in that. Because quite a lot of SSIS packages I've seen deal with data cleansing. If that can be automated via off-the-shelf packages, then people would probably prefer that compared to doing it by setting up packages.

Julie: Yeah, a data lake is another methodology that is going to change data integration. So maybe we'll be using SSIS just to move files to a data lake and do less of the hardcore transforming inside SSIS. So, you might still use it. We might just not use it to do the transformation work.

Mala: Interesting.

Julie: We're always going to need batch tools to move data around.

Mala: Right, right. Describe a few things that a BI professional should know as best practices.

Julie: There are a few things I want people to know as a BI professional. I want you to understand the Kimball methodology. I want you to be familiar with data lake methodologies. And, at this point, I do think that people need to be very familiar with analytics technologies. When I say "analytics," there are two ways to interpret it. You've got predictive analytics, which is machine learning and artificial intelligence. But "analytics" also refers to visualizing data. Like the visualizations in Power BI or Tableau.

Mala: One of the much sought-after skills now is data visualization. What has been your approach to this and how do you recommend learning more about it?

Julie: Great question. A project I'm working on now was my first foray into using visualization. There are tons of resources available online—courses, blogs. There are books. Coming to the project with a history of data integration, I found the amount of information about visualization overwhelming. I focused on the basics.

I watched Dan Appleman's Data Visualization for Developers course on Pluralsight. I found a few "pillars" that I could cling to there. I mainly used four main types of visualizations based on the recommendations in the course: bar or column, line, scatter, and tables. Lines are great for one value that changes over time. Scatter is used when the data over time has outlier tendencies that are important to see. Tables are helpful to communicate actual numeric figures. And bars or columns are great for almost everything else. If you start there, you can get fancy later, but these are great basics. Then just practice, practice, practice.

Mala: Simple stuff, yeah. What are some recommended best practices in the BI world that you like and recommend?

Julie: So, I feel like our best practices are changing. There are people out there showing us how to use the technology. For Power BI, Guy in a Cube is amazing because they give you short videos on how to do this. By the way, that's how I learned SSIS— from Brian Knight's videos. They're not even available anymore. I don't know what they're called, but Brian Knight used to have these ten-minute videos. This is back in like 2007. I watched those, and I feel like that's the best way to learn, is just say to yourself, "I need to do this. I figured out a business problem, now I need to do it in the technology. I need to know how a derived column works. Oh look, Brian Knight's going to show me how to use a derived column." That's what Guy in a Cube, Patrick LeBlanc, and Adam Saxton do now with elements of Power BI.

That's learning the technology, but you asked about best practices, and I have joked that I don't have a good handle on best practices in this day and age because the technology is changing so fast. But I want to mention that Melissa Coates—SQL Chick—is amazing. She publishes guides on her blog that outline best practices for Azure and Power BI. I'm so grateful to her for publishing them. I'm starting to adopt her best practices in my work. The most important thing is to be consistent.

Mala: What is the role that documentation and communication play in your job?

Julie: Communicating what I am doing for a client is as important as the actual artifact that I produce. Documentation plays a role in my overall communication strategy. I focus on the basics. I'm engaged by my client to fix business problems. If I need to meet with you face to face in order to gather the requirements for a report, I will do that.

Back to what I said previously. Communicate frequently. "This is what I've done. Does this work? If it doesn't work, what is wrong? I will fix that by this date." Dates and tasks should be in writing so that there is no confusion.

Mala: Do you want to elaborate more on how you have seen agile work with data warehousing?

Julie: I have used agile with at least two clients. So, I've been through a formal agile with scrums and backlog grooming. I think it works well, and I think it's pervasive, so I think everyone needs to know agile and be able to work with it. You wind up with the same problems in agile that you wind up with in any software development life cycle process. You can mess up in agile just like in waterfall.

I think that agile really does need the business to be involved. I've seen that overlooked frequently, where the business consumers of the product aren't adequately involved. It winds up being just the tech people in the scrums. That is my big advice to everyone: frequently check in with the end user of the product that you're building. Make sure that you didn't misunderstand. Ask. Show them the work in progress. Don't assume.

Mala: Makes sense.

Julie: Tech people prefer to talk to other tech people sometimes, and it can be a challenge to stretch and have a talk with the business.

Mala: Oh, yeah. It's easy. It's much harder talking to the business because you have to step out of the job and get very down-to-earth.

Julie: It's nerve-racking. They're often the ones who are going to make or break you, and so it can be nerve-racking. Human tendency is to avoid things that make us uncomfortable, but do it because it will help you succeed rather than fail. And that's how you gain trust, and that's how you get more business and more trust.

Mala: Describe your experience with cloud adoption.

Julie: As I've said, I've been on a client since July 2015. They are in the construction industry—home building. During the recession of 2008, they suffered, and so when things started to recover, in 2015, they said, "We are never going to have a server room. We're never going to have IT staff." They wanted to use all cloud technology. So, they've been amazing. They loved using new technologies, so they've been good for me in that we are all Azure. The thing that is interesting is that we started the way most people start, where it's just the VM (virtual machine). So it's really like you're in the on-premises world, only they're VMs in Azure. But we were able to encourage them to use the platform as a service, and of course, Power BI to get them into software as a service.

So, for the last two and a half years, I've been able to do tons of cloud and get lots of Azure experience. Hands-on experience. I've used Azure Search, Azure Data Factory, and Azure SQL Database. The client's applications are built entirely with Azure Services. So, I have had good luck with that, and it has worked beautifully for them.

Mala: Wow.

Julie: I'm interested in what other people are finding. I mean, I see it. About two years ago, I saw adoption start to snowball. It's definitely the way things are going. There's no escaping it.

Mala: Most people I talk to prefer a blended approach to it. Very few are one hundred percent into the cloud. It's sort of evolving slowly, I think. I think that's the way it should be.

Julie: It definitely should be if you are an existing organization with on-premises resources.

Mala: Oh, that is correct, yes.

Julie: I mean, there's no way that you should say, "We have to be a cloud, so we're going to migrate everything overnight." No way. The only way that I think an "all cloud" approach works, and I've seen this, is if it's a brand-new organization. Then you can do it.

Mala: Yeah, you don't have to invest in a data center or anything like that, so, yeah, sure.

Julie: You can replicate on-premises to an Azure SQL Database and vice versa, so it gives you disaster recovery options as well.

Mala: What are some of your favorite tools and techniques?

Julie: Tools and techniques. I am an integration person, so my hammers are SSIS and TSQL [Transact-SQL]. I do things really quickly in integration services. I'm a fan of Biml [Business Intelligence Markup Language]. I have little tricks that I use for data integration. I like to use the row number function and the partition by.

Mala: Oh, I love that, too.

Julie: Yeah, that is really one of my favorites. I use that with the business key too. In warehousing, often you have to de-duplicate. So with the rownumber function, you can partition with the business key, and sort with the tiebreaker, which groups the rows that are dupes and numbers them so you can grab and load just one of the rows.

Mala: Makes sense.

Julie: Yeah. That's one of my favorite tricks. For doing data integration testing, I use a union, intersect, and except. There's a lot of data duplication—like I said, the drudgery work. So, I use T-SQL. Oh, common table expressions. I don't know what I would do without them.

Mala: So true.

Julie: Yes. Let's see. Yeah, I think that's off the top of my head.

Mala: That's fine. What are your favorite books, blogs, and other means of learning and keeping up?

Julie: You have to follow whoever and whatever platform you're on. I work on the Microsoft platform, so I follow a lot of the Microsoft blogs and principals there. You have to follow blogs. You have to follow folks on Twitter. Like I said, I got started with Brian Knight's videos. Guy in a Cube is essential for Power BI. Melissa Coates for best practices on emerging technology. It's funny, I am in the community, so a lot of the time, you get to know somebody personally and then find out that they're just an amazing resource.

Mala: So true, yeah.

Julie: Yeah. For my forays into big data, I look to Michelle Ufford. She has a blog called hadoopsie. When she was mostly a SQL Server database administrator, it was SQL Fool, but now she's in the big data world, and so it's a Hadoop site.

There are people that are foundational, so Jamie Thomson for SSIS. I try hard to keep up with the industry as a whole and Jen Underwood's blog is amazing for market analysis of data products.

But Twitter, you set up TweetDeck, and then you create columns for the products you work with. I have a column for Power BI because that's the one that I watch the most. It's changing literally weekly. Then when something comes up that you've been waiting for, you just say, "Oh, okay, it's on Twitter," and then there's a link and you follow it. You don't have to learn it right then and there. All you have to do is know that it exists and you know where to find it.

Mala: Correct. So true. What are your recommended ways of stress management and developing a healthy work/life balance?

Julie: That's hard, and I credit my children with helping me with this because kids have a way of forcing work/life balance. When your kid needs something, you can't say no. "I have to go get my kid. My kid is sick at school, so I can't be here." I find that I struggle with it. I take yoga classes when I can, and I do a short yoga routine every day at home with some apps I found.

I find that I tend to get frazzled, and that's when I'll realize it's been three or four months since I took any time off. I get to the point where I can't remember anyone's name, and I feel like I'm made of lead. And I'll look and realize, "Okay, it's been a while since I took a break," and then I just have some off time.

I do manage to regroup during downtime. I just had some off time because it's the Christmas holiday right now, and I worked on my journal to plan and regroup. During my downtime, I like to just sleep and lay around, and then I feel my motivation returning. You have to take time away, completely off. You have to take at least a day or two where you just don't do anything. The Oatmeal has a cartoon about it. It's breathing in. Our work is exhaling, and you have to have a time where you're allowed to inhale.

Mala: Yeah. That's a simple but very true statement. Describe your style of interviewing. What do you look for in people you interview? What are the types of questions you ask?

Julie: I ask a few questions that are technical. I try to find a way to determine if there's that problem-solving ability in a person, because I feel that's the most important quality a developer can have. So, I ask questions like, "Tell me how you solved this problem. Tell me a story about a project that you did and why you did it." You can hear just in the way that the answer is given how the person thinks and if they're worried about impressing you technically or just thinking about how to solve business problems. I look for the latter. Having said that, I would say that it's amazing how, honestly, I feel, ineffective interviews are. That is why your reputation, word of mouth, and references are so much more important.

Mala: Oh, I agree completely with that.

Chapter 2 | Julie Smith: *Business Intelligence Consultant, Innovative Architects Inc.*

Mala: Right, right. What are your contributions to the community? Why do you recommend people be involved with the community?

Julie: I have been speaking and presenting in the community since 2009. I have a blog, The Datachix, that I share with Audrey Hammonds. One of the things that makes me so happy, other than making people laugh, is when someone thanks you for something that you've helped them with.

When SQL Server 2012 came out, there was a big change for SSIS from the previous versions, and I had a blog series about how to make that transition. I've had people say that they used my blog series as a way to learn it. Or when someone comments on a blog post that I saved their life—figuratively—with the information in it. It's very rewarding. It'll get me through a rough day otherwise.

If you're having a rough day, especially at work, you get down on yourself. You say, "Well, maybe I'm not very good at this data stuff." But you can say, "Okay, well, I'm helping people. I really am good at this."

So, I blog and I speak. And why do I recommend people be involved with the community? It's always been so helpful to have resources at your fingertips in which you can tweet a question with the hashtag #sqlhelp or #powerbi, and you likely get a response from literally the expert in the field. So being in the community gives you unbelievable resources. I would say it's almost essential. I don't know how a person could succeed in data right now without being involved in the community, at least being aware of the community and utilizing the resources.

You can utilize the resources, but you will get more back if you also pay it forward. That doesn't have to be speaking. You can volunteer at events and help. You can always just write. Even if it's not a blog post, you can help someone on Twitter or in forums. You can be a bit of help. So, it's just essential. I would advise it. In any of my presentations now, I'm like, "Get on Twitter. Even if it's just so that you can follow others."

Mala: This is the last question. If you had one superpower, what would it be and why? It does not have to be work related.

Julie: Oh, my goodness. Superpower. I wish I had instant hindsight. I guess that is the superpower of seeing the future. Oh, I look at a picture of myself. When it was taken, I was so worried about my job and my appearance. In hindsight, the situation worked out—and I realize now that I look just fine in that picture. But at the time, no one could have convinced me of that.

Mala: Let little things be little things.

Julie: Right, and that's the work/life balance. I think I am better at that now—not getting so ramped up.

Mala: Right, so true. So true. Thank you, Julie!

Key Takeaways

- Do things manually first and then figure out how to automate it.

- Involve the business in agile. Frequently check in with the end user of the product that you're building. Make sure that you didn't misunderstand. Ask questions. Show the work in progress. Don't assume.

- Get dates and tasks in writing.

- Build a good reputation and references.

Favorite tools: Biml, SSIS, T-SQL

Recommended training: Dan Appleman's Data Visualization for Developers (Pluralsight), Guy in a Cube videos for Power BI

Blogs to follow: Melissa Coates' SQLChick.com, Microsoft blogs, Jamie Thomson's SQL blog.com, Michelle Ufford's hadoopsie.com

Kenneth Fisher

Senior Database Administrator, Aegon

Kenneth Fisher has spent his entire 30-year career working with data in one form or another. He's used SQL Server since SQL 2000, and he has enjoyed almost every minute of it. Kenneth is a strong advocate for blogging. He has blogged twice a week at SQLStudies.com for more than five years. He keeps a public list of blogging ideas in case anyone else needs one, as well as lists of free tools and education resources. He also believes that learning should be fun—he writes SQL Server–related crossword puzzles for SQLServerCentral.com. He was awarded the Microsoft MVP in the Data Platform category.

Currently, Kenneth works at a large international insurance company with a couple of dozen other database administrators (DBAs). His job is best described as "supporting the developers."

Kenneth's current hobby—aside from SQL Server—is making small metal models. He enjoys spending time with his wife of 20 years and his 14-year-old son and 8-year-old daughter. You can find him in person in Sherman, Texas, or on Twitter as @sqlstudent144.

© Malathi Mahadevan 2018
M. Mahadevan, *Data Professionals at Work*,
https://doi.org/10.1007/978-1-4842-3967-4_3

Mala Mahadevan: Describe your journey into the data profession.

Kenneth Fisher: Okay, I'm going to start with one word. Nepotism. I met this guy when I was in college. We were taking a programming class and became good friends. Later, he ended up getting me my first job and several other jobs as well. The big one was where I was living in Houston and wanted to move back to Austin. My friend told me, "We need a new DBA/developer for FoxPro. I said, "I don't know FoxPro." He said. "That's okay. I've told the boss that you have a year experience in it. I'll help you wherever you get stuck." It helped that he was the hiring manager, too. So needless to say, I went to work there. He had a very interesting style of helping me. This was twenty-five years ago. There was no Internet as we know it today. His method of helping me was to tell me to look in books and at the online help. It was frustrating but it taught me to be very self-reliant. From there, I ended up with another job with him. I was doing FoxPro, and they started converting to SQL Server. I decided at that point to give it a shot as a contractor for a while.

I quickly realized I hated being a contractor and went back to work at the same company that I had just left. I was doing FoxPro to SQL conversion, which I quite enjoyed. From there, I went on to several other SQL jobs. Years later, I went to my first conference, which was the PASS rally in Dallas. I loved it. I fell in love with the community. I fell in love with the sessions, the people, and the energy. A few years later, I started blogging. Over time, I wound up here.

Mala: Describe a few things that you wish you knew when you started your career that you know now and that you would recommend to newcomers in this line of work.

Kenneth: Don't assume you know everything. You don't. The more you learn, the more you're going to realize you don't know. Be willing to learn. Always be involved in learning something new. Share what you know because that's the best way to learn it, at least in my opinion. Enjoy the community. We're very lucky to have a very active community. I've talked to a guy who had a database get corrupted at three o'clock in the morning. He got online and there were people in the community who happily spent the rest of the night talking him through how to fix it. The more you put into the community, the happier they're going to be to help you through situations like that.

Mala: How can people put more into the community?

Kenneth: Answering forums is a great way to learn stuff.

Mala: Give some examples of forums that you like.

Kenneth: I'm very big on Stack Exchange, the DBA part of Stack Overflow. SQL Server Central is also really good, as is Ask SQL Server Central. Those are the ones that I primarily use, though there are others out there. By answering a question, you'll learn a lot because you have to condense your thoughts and your knowledge down to something that somebody who's just reading this—who isn't in front of you to ask questions—is able to understand. You have to recognize the little pieces that don't really make sense. You have to go out and learn what needs to go into filling that hole. You'll format your knowledge better. You'll learn it better. You don't necessarily get a whole lot more depth, though you can. It will solidify the knowledge you have a bit better.

There's blogging, of course, which is what I tend to do a lot. It's similar to answering questions on a forum. You have to put your thoughts down in such a way that you're teaching someone who isn't there to ask questions. You really have to format it well, so that they're going to be able to read it and make sense of what you're doing.

Like right now, I've been thinking about the fact that in Azure you can create—by using one of their templates—a VM with SQL Server already on it. But you have very limited ability to change the settings. I tried it, and I wrote a blog about it, just to see what the limitations were.

Now I'm doing the next step, which is to create a VM and install SQL Server separately. From there, I'm going to try and create a template that creates that VM with SQL Server with those settings on a third VM. Each one of those is going to be a blog that describes how it went. It may not work out. Sometimes the answer is, "No, you can't do it that way." Sometimes that's worth writing in a blog and sometimes it's not. Either way, you learn something.

There are so many different ways that you can become part of the community. You don't even have to share. You don't have to be the type that blogs, or speaks, or answers forum questions. You can be the kind of person that sets up SQLSaturdays or that goes to SQLSaturdays and fills bags, and hands out stuff and directs traffic. They're all necessary pieces to making this stuff work. We keep talking about SQLSaturdays because they're very common, they're free, and they're easy to do, but they're just one of several different community events.

There's also just your local user group—PASS or something else. And online users' groups. They're always looking for speakers. They're always looking for help with things. There's a lot of work that goes on behind the scenes. Even if you choose not to do any of that, going to the sessions and just being a part of it in the sense that you're a body filling a chair, you're learning and it's worthwhile.

Mala: What is a typical day in your life as a DBA?

Kenneth: I like answering questions. I enjoy that. Problem solving is one of the things that I really have fun with. So that's where I've directed my career. A day in my life is getting the obnoxious problems. I get the person who's having a hard time connecting or needs help tuning a difficult query. I get a lot of calls from co-workers who are also DBAs.

My company is a very large company. We have more than thirty DBAs who do approximately what I do, though not always with SQL Server. But fifteen to twenty of us are, at least to some degree, working on SQL Server. That doesn't include the hundreds of developers and the thousands of programs that we work with.

There's a definite need for somebody who doesn't do much else other than answer questions. Now, I also get the fun projects—at least, I consider them fun. One I'm doing right now is converting a database from Oracle to SQL Server. It's not exactly complicated, other than being able to connect to Oracle. I'm just going to use SISS [*SQL Server Integration Services*], but I've got to get the connection to Oracle to work right, and I've never done that before. I tend to get those kinds of things.

I've also gotten to do some cloud stuff. Both AWS [Amazon Web Services] and Azure. Maybe not as much as I probably should, but I'm getting there. I'm one of those people who is pretty specialized. I know some administration stuff. I know some development stuff. I don't know much in the way of BI [business intelligence]. I do know some ETL [*extract, transform, load*]—things like that. But all SQL Server. My day is spent working on various projects and being interrupted with questions and answering questions. "How do I take this backup? What happens if I do that?" That kind of thing.

Mala: What industry trends are you typically watching out for?

Kenneth: One of the things I've noticed is that there's a certain amount of following the leader in this. I look at what the industry leads are doing. Personally, I'm looking into the cloud. I don't know that I agree that you can't just know SQL Server. At least to a certain extent. I mean, you couldn't only know SQL Server in the past. You had to know Windows—to a certain extent—just to function. Now you have to know the cloud, as well, just because it's another environment that you're going to be working in.

The more you know about these things, the better off you are. Knowing how to RDP [Remote Desktop Protocol], knowing how to use Citrix, knowing how to use all these other communication tools is very helpful. Brent Ozar just did his yearly survey on how much money DBAs make. One of the things he broke out was male and female pay. One of the things I noticed while he was doing that was that the generalists, the people who do some administration and some scripting, tend to make less than those who are only developers or only administrators. It's not a big difference, but it does tend to make me think that there is no shortage of the need to specialize. Look at people that only tune databases, or tune queries, and tune instances—those are important skills that will get you a job. People who specialize in high availability, which is an important skill, will be able to find a job.

Mala: Now describe a few things that a DBA should keep in mind as best practices.

Kenneth: My main best practice is a meta best practice. Don't believe a best practice just because it's a best practice. If you've read that setting the max degree of parallelism to eight is the best practice, why is it the best practice? Because tomorrow it may not be. The best practice for years was to make sure that you had your log file on its own drive, that you use multiple file groups, and that you split your data across multiple file groups. That way, you have multiple spindles hitting your data and you get more throughput. Yet, now with SAN [storage area network] technology, and SSDs [Solid State Drives] and all that, that's not as necessary anymore. You've got to know when the best practice is no longer the best practice. The only way you're going to know that is if you know why it's the best practice.

Mala: Correct. Very well said.

Kenneth: A second important best practice is to always be recoverable. When I first began to program, I learned to save and save often. If your program crashes, or if the computer crashes, you want to be able to get back to where you were. It's the same thing with databases.

Mala: Be recoverable is very wise advice. What are a few things that you see as worst practices? Typically, things that juniors do— maybe where you are— or things that people who are starting out should look out for?

Kenneth: I hate NOLOCK. All the things a senior DBA hates—NOLOCK, cursors, whatever else. They are useful tools, but senior DBAs hate them because juniors use them everywhere, including in places where they shouldn't be using them.

My worst practice is when you stop questioning what you're doing. You're just doing it by rote.

Mala: Very good advice. Do you have any experience with using agile in your workplace?

Kenneth: I do. I work at a very large company with lots of teams of developers. It is not possible for my company to do agile in the sense that a DBA is part of every team. We would quite literally be in scrums eight hours a day every day, and miss a number of them. We'd never get any work done because we'd be in scrums all day. I think the agile concepts are fine. I think they're a good idea in general. I don't think they work overly well for DBAs as part of the team.

Mala: Do you generally feel that it's made you more productive than with the waterfall model?

Kenneth: For me, no, not particularly. Again, in the companies that I've worked with, it was the developers who were affected by the waterfall model versus the agile model.

Mala: You do more frequent deployments, so your cycle of work is somewhat impacted by that at least.

Kenneth: Well, again, I'm working for a company with dozens upon dozens of teams. When it was waterfall, we would still get deployments from different teams. They were just bigger deployments. One team would have a huge deployment over the course of a couple of days, and the next team would have a big one over the course of the next couple of days. They did tend to spread out to a certain extent. Whereas now, it's multiple teams having smaller deployments. I don't think it's really affected me much with this company.

Mala: What are some of your favorite tools or techniques?

Kenneth: My favorite tool is SQL Server Management Studio. I use it to do text editing because I'm so familiar with it. I love the using Alt key, for example.

Mala: Are there any monitoring tools or any DBA monitoring tools that you prefer? Or do you like homegrown stuff—like third-party tools?

Kenneth: I tend to write my own tools. But I really like Minion. I don't get to use it for work because I have no control over that aspect of things. Also, sp_whoisactive is a very useful tool. As far as monitoring tools go, I like IDERA's SQL Diagnostic Manager.

I'm slowly but surely learning PowerShell. It is not yet my language of choice by any means. It is one of those things that you need to know. One of the things that you need to know, no matter what your specialty, is at least some PowerShell.

Mala: I totally agree.

Kenneth: Just because it's so ubiquitous these days. I think it's great. I grew up with Command Shell, with DOS. The idea that this PowerShell thing has come along that basically replaces that and does far more than the Command Shell ever could, is just amazing to me. That's another one of those things that I'm trying to learn over time. I have a script similar to sp_whoisactive. I've got several index scripts. I have a backup script that checks where backups are being taken and have the template for creating a backup and restoring a backup. Things like that. The solution is useful because it keeps everything organized.

Mala: Makes sense. What are your favorite books, blogs, and other means of learning?

Kenneth: Well, honestly mine is probably the place I go to most often, but that's on purpose. I find that if I've written about it, then I can go back and review it. I review what I wrote as a quick reminder of a subject. It'll get me a jump-start. Beyond that, SQLskills—their stats stuff is amazing. I couldn't live without it. I like Aaron Bertrand's stuff. I'll read Denny Cherry's blog[1] occasionally. Brent Ozar's blog[2] occasionally. I'm very eclectic. I'll pretty much

---

[1]https://www.dcac.co/publications/blog
[2]https://www.brentozar.com/blog/

pick whatever passes me randomly at some point and kind of skim it. As far as books go, there's an SSIS book I really like, which is Knight's *Microsoft SQL Server 2012 Integration Services 24-Hour Trainer* [by Brian Knight, et al., Wrox, 2012].

Mala: I like those books too.

Kenneth: I also like Learn PowerShell in a Month of Lunches [by Donald W. Jones and Jeffrey Hicks, Manning Publications, 2016].

Mala: What are your recommended ways of stress management and developing a healthy work/life balance? Generally, the DBA profession is regarded as stressful by a lot of people, although it's very rewarding. What do you do to take your mind off of work?

Kenneth: I did an email interview with Thomas LaRock years ago, and I asked him about work/life balance. He had an interesting perspective, and he said there really isn't a work/life balance. You have work and you have life, and really, it's the same thing. Work is just part of your life. You just can't let any part of it get overwhelmed. I don't remember exactly how he put it, but basically, the idea was you can't look at them as separate.

As far as stress goes, my personal opinion is find something that is outside your stressor area. If thinking about stuff is what stresses you out, find something to do that doesn't require a lot of thought. I build models. I do these little metal models, and I keep them on my desk. If I'm feeling a little stressed out or I need a break, I'll spend five minutes, and I'll put another piece on.

Mala: Oh, wow. Okay.

Kenneth: Then I'll set it aside, and I'll get back to work. It requires concentration. These things are very small. I use tweezers, pliers, and things like that to manipulate them. But it doesn't require thought if that makes sense. I get a break for a few minutes, and then I come back. I also find blogging to be very relaxing.

Mala: Yeah, me too.

Kenneth: Find something that works for you. It's different for everybody. You look at several of the people we both know online, and their big stress relief is jogging. I am not a jogging person. That would not work for me.

I know a lot of people who color.

Mala: I do. I'm one of them too.

Describe your style of interviewing a DBA. What do you look for? What are some examples of questions that you ask?

Fisher: The questions I ask depend on what I'm interviewing for, obviously, and the level I'm interviewing. If I'm interviewing a junior who's going to be working with the administration, and who's not strictly a developer, I might ask about system databases. I might also ask the odd question about queries. For someone who's mid-level, I'd probably ask some of the same questions. One of my favorite questions tells me about clustered indexes. If you're a junior and you know the clustered index is the table, that's sufficient. I want a senior to tell me in more detail, and to explain what effect a clustered index has, and so on and so forth.

That said, what I'm looking for, aside from a basic level of knowledge that matches the job requirement, is enthusiasm. I want somebody who wants to be there, and who's excited about the work. My favorite interviews, on both ends of the table, have been ones where we pass trivia back and forth. Not that everybody likes this, but I happen to like trivia. It doesn't matter if you know the answer. If I say, "What happens if you run a log backup at the same time as you run a full backup?" You don't have to know the answer, but I want you to be interested in the answer. I want you to say, "You know, I have no idea. What happens?" Or, "That's really cool. I haven't thought about that. I'm going to have to go look at that when I get home." Then I'll share the answer.

If I'm looking at a senior-level person, once you get past a certain level of necessary knowledge, I want somebody who's excited about the job—interested, willing to learn, having fun, and who will walk in in the morning excited about being at work. Maybe not every day, but in general likes their job. If they're struggling to come in, or it's just work, it's boring, and they don't really want to be there, then they're not going to do as good a job in my opinion. I can teach people stuff. If they want to learn and they want to do the work, then I can teach them. I have no problem with that, especially in the case of a junior level or something. If somebody's excited about it, I can teach them. I can't teach excitement.

Mala: Very true.

I know you really like certification, so can you share some thoughts on what you feel about people getting certified? Do you think it's necessary?

Kenneth: I think certifications are very, very helpful tools if done in the right way. If you're using a certification as a framework to learn, then the certification itself is just a bar. "Did I learn at least a reasonable amount of this material? Do I need to go back and start again?" Then absolutely, I think they're an amazing thing. They do give you that format. They give you that classroom format that we've been trained into. I would have never touched Service Broker had it not been for a certification. I am not an expert in Service Broker, but I know what it is. I know what a queue is and I know the basics. If somebody ever says, "I need to be able to do asynchronous messaging." I'm going to reply, "Service Broker." I'm going to look up the commands, and I'm going to know to use Service Broker. But I'd have never heard of it before if it weren't for a certification.

That said, if you are the type of person who wants an answer dump, and you just want to memorize the questions and the answers so that you can take the exam, but you never really learned anything, then no, just don't bother. It won't help. Honestly, if you walk into an interview and say, "I have a certification," but you can't take a backup, you're less likely to get the job than if you walk into the interview and you don't have a certification and can't take a backup.

Mala: Correct. Well said.

Kenneth: If you're learning the material, awesome. If you're not learning the material, certifications are worse than useless.

Mala: If you were given one super power, what would you choose and why? It doesn't have to be work related.

Kenneth: Do you remember the TV show Heroes?

Mala: Yeah, but I don't think I've seen it. Go ahead.

Kenneth: It was a great show, but they messed it up by bringing in two overpowered people. They both had variations on the same ability. They could use other people's abilities. One of them actually stole their ability and the other one just duplicated it. That second one. The ability to duplicate what other people can do.

Mala: Interesting.

Kenneth: That would be awesome. I see someone do a cartwheel, now I can do a cartwheel.

Mala: That would be fun.

Kenneth: I see someone run a three-minute mile. Okay, I can now run a three-minute mile. I watch someone take a test, and now I can be that smart.

Mala: I'd like that too. That would be fun. Thanks, Ken!

Key Takeaways

- Be involved in the community. Choose what works for you—blogging, speaking at events, Twitter, answering questions on forums, or any other means.

- Question best practices and re-examine recommendations before following them.

- Follow industry leads to understand where and what technologies are evolving.

- Show excitement and be passionate about what you do.

Favorite tools: SQL Server Management Studio (SSMS), Minion, Adam Machanic's sp_whoisactive, IDERA's SQL Diagnostic Manager

Recommended books: *Learn Windows PowerShell in a Month of Lunches* by Donald Jones and Jeffrey Hicks, *Knight's Microsoft SQL Server 2012 Integration Services 24-Hour Trainer* by Brian Knight, et al.

Recommended training/conferences: SQLSaturday, PASS Summit

Andy Leonard

Chief Data Engineer, Enterprise Data and Analytics

Andy Leonard *is the founder and Chief Data Engineer at Enterprise Data & Analytics, creator of the DILM (Data Integration Lifecycle Management) Suite (*dilmsuite.com*), an SQL Server Integration Services (SSIS) trainer, consultant, a Business Intelligence Markup Language (Biml) developer and BimlHero, SQL Server database and data warehouse developer, community mentor, engineer, and farmer. He is a coauthor of The Biml Book (Apress, 2017) and SQL Server Integration Services Design Patterns (Apress, 2016), and the author of Managing Geeks: A Journey of Leading by Doing (CreateSpace, 2014), Data Integration Lifecycle Management with SSIS (Apress, 2017), Building Custom Tasks for SSIS (Apress, 2017), and the "Stairway to Integration Services" articles at SQLServerCentral.com. Andy blogs at AndyLeonard.blog. He may be followed on Twitter @AndyLeonard.*

Mala Mahadevan: Describe your journey into the data profession.

Andy Leonard: I was a software developer years ago. I actually started working with software in 1975 when I was eleven. My neighbor retired from the US Air Force, and he returned home to take care of his aging parents.

© Malathi Mahadevan 2018
M. Mahadevan, *Data Professionals at Work*,
https://doi.org/10.1007/978-1-4842-3967-4_4

I used to hang out at his house and watch him do engineering stuff. He actually worked for the Air Force on a component that kind of worked with NASA. So, a pretty sharp engineer. He built a computer from a Southwest Technical Products kit. I learned Motorola 6800 machine code from him in the summer of 1975. And then that Fall, I learned BASIC.

Computers were a hobby for about the next twenty years. I became an electrician and worked in a plant. I worked evenings and nights while I went to school. The day engineers were trying to solve a problem on a rotating machine that created hardware, and they were programming Allen-Bradley programmable logic controllers. It turns out that those are like the great, great, great grandchildren of the Motorola 6800. They used the Motorola 68000 series controllers. We could program the 68000s in machine language. I remembered enough machine language to help solve that problem. It kind of crossed a barrier between being a hobby and being a job. I got into programmable logic controllers—or PLCs. I began working with human-machine interface—or HMI—industrial software. It began blending about the time the Web came along. I think I got on CompuServe around 1993 or 1994.

Later, I landed a gig as a software developer in a manufacturing plant. One day after I left that job, the plant manager called. He asked, "Can you build a database application that would give us a report on how many times this machine stops, when it stops, and how long it stops?" Basically, an uptime-downtime processing app. And he said, "We need a database application for this. Can you do this?" Now this was back in the mid-nineties. I went back home, logged into CompuServe, opened Netscape, and navigated to altavista.digitial.com. This was before Google. I typed the word "database" and hit Enter. I did not know what a database was. But I told the plant manager I could do it.

Mala: Oh, wow!

Andy: Yep. I sold that first database application for five grand. I wrote it in Access 2.0 and Visual Basic 3, and interfaced with a National Instruments card on the ISA bus of a PC used on the manufacturing line on this machine, and it worked. I built it. It's pretty neat. That was when I really got into doing things on my own. I didn't know what a database was until then.

A couple years later, I was working with human-machine interface software for a plant. It was called Wonderware, and it's still out there. Wonderware uses data acquisition points that they call tags. To stress test the system, I set all of the tags to their highest frequency—record once per second. I used Access to store the data. I turned it on for a long weekend. I returned to the plant on Tuesday and there was this three-gigabyte Access file. It wouldn't open.

Again, I went to my favorite search engine and entered "Microsoft database." I saw an entry for something called Microsoft SQL Server 6.5, and I bought a not-for-resale copy at a computer show about two weeks later. I took it to the plant, installed it, and reran the test. It was able to store the three gigabytes of data. I told the plant engineer what I'd done. It was a good proof of concept. Microsoft SQL Server 7.0 was just coming out at that time. They bought a valid license for SQL Server 7.0 based on my recommendation. That's how I first started using SQL Server.

Mala: Wow. What a story. One of the best stories I've heard.

Andy: Thanks.

Mala: Describe a few things you wish you knew when you started and would now recommend to people who are new in this profession.

Andy: Two things leap to mind. One is that failure is normal. The second thing kind of goes along with that. Imposter syndrome is also normal.

To elaborate on failure, I had all these emotions attached to whatever I built, especially if it didn't work. I felt that not just my code had failed or what I was trying to do had failed. I felt like I had personally failed. I'm competitive and pretty ambitious, but it's not directed outwardly. I'm trying to beat "Andy from Yesterday." When I would fail, I felt like I didn't beat Yesterday Andy. I felt like I lost the competition against Andy from Yesterday. That was something I battled regularly during my early career.

That fed the imposter syndrome. So whenever I had a job, whether I was contracting or working for an employer, I always felt like any day they're going to figure out I don't know what I'm doing, and that I fail all the time. And then they were going to fire me.

It took years for me to start realizing "imposter syndrome" was a real thing. I learned this through the community, really. The SQL Server community helped me a lot with this. You all helped me understand that failure is not only normal, it's actually good. It's a part of the success. The pathway to success. I'll give Andy Warren kudos and credit for helping me with this. Also, Steve Jones. They write a lot about this.

Mala: Correct.

Andy: They really help me see. "This is not only normal, it's good." They wrote about imposter syndrome, and I heard some talks at SQLSaturday and at the PASS Summit. I realized I'm not so bad off.

My very first PASS Summit was in 2004 in Orlando. I heard two of my heroes, Kalen Delaney and Ken Henderson. Kalen, of course, was awesome. She's still awesome and still one of my heroes. Ken Henderson is no longer with us.

Mala: Right.

Andy: I had read Ken's books, and I was in his session. After that, they had something like a meet-the-experts time. I was in line getting ready to ask him some questions. I was working on my very first data warehouse. I had a lot of questions. You'll find this funny. There were no surrogate keys in this warehouse. It was all natural keys. I didn't design it, but I didn't know any better at the time.

I'm sitting there waiting in line. I'm basically hacking at this thing using my engineering background. You know, just experience, mostly stubbornness, which is a virtue, I think! And this guy in front of me said something to Ken and argued with him. And Ken just cut him off at the ankles. Then it was my turn. I go up to Ken and I'm like, "I don't really know what I'm doing. I'm a software developer. I've just started doing database work professionally, and I've got a lot to learn, but here are the problems I am trying to solve." He gave me some really patient advice. I could see his face was red from talking to the last guy. He kind of calmed down talking with me. He gave me some great pointers. I was writing as fast as I could. At the end of it, I thanked him profusely. I'm walking away, and he says "Hey." And I stop and look back, and he says, "You know, you're approaching this problem like I would." I got to tell you, Mala, I walked into that room the biggest hacker and fake on the planet, and I left that room a data professional.

Man, it was great. I wish I'd known more about the fact that failure is—just take the emotion out of it and realize that you just learned some stuff, so that next time you won't repeat that mistake. It's part of the process.

We're all learning. It's normal to feel like other people know more than you. Imposter syndrome is normal. You're better than you think you are. That's what I'd share with others. You really are. You know more than you think.

Mala: What's a typical day in your life as a professional?

Andy: I am a small business owner. I own one company that does business as really three others. A couple of those businesses have actually more than one venture going on at a time. I usually wake up around five o'clock. Sometimes I'll sleep in or hit the snooze button a couple of times and get up around five thirty. About half the time, I actually wake up before the alarm goes off, sometimes as early as three o'clock or three thirty in the morning.

Mala: Wow. Oh, my.

Andy: I am old, Mala. Old people don't need much sleep! Sometimes I'll try to go back to sleep but after about five minutes, I kind of know, "Nope, not going to happen!" I'll get up and usually fix a cup of tea, head up to the office. It depends on what kind of mood I'm in. Here, lately, I've been splitting my start activity between taking training courses. I've mostly been taking Microsoft Virtual Academy courses that are offered in cooperation with edX. I picked up a few certifications. I'm working toward the big data program certificate, which is nine courses plus the capstone. I'm a little over halfway through that program. A lot of the courses overlap with the data science certification, so I'll probably complete the big data certificate first.

Hopefully, I'll finish off the data science program late this year or early next. Along the way, I've really picked up a lot there. About half the time, I'll do that two or three days a week. I'll do that in the morning. The other thing I may do early in the morning is work on SSIS Catalog Compare, which is a product. It's a part of the Data Integration Lifecycle Management Suite. DILM Suite. A little shameless plug there.

I built a bunch of free utilities. Catalog Compare is the anchor product that I sell. It also serves as an anchor for another product called Catalog Browser, which is free. I surface everything in the SSIS Catalog in a tree view and try to make it so you don't have to open a bunch of dialogs and windows.

If you've worked with the SSIS Catalog in the SSMS [SQL Server Management Studio] node, there are a lot of windows that you have to open up to see configuration information. It can be confusing, I think.

The SSIS Catalog is really good the way they built it. That surface is confusing, and I'm trying to solve it. I don't know if I'm achieving that or not, by putting everything in a single tree where you can just expand things out. The version I'm working on now, I'd originally hoped to have out in October. You and I are talking in January. Then I thought I'd have it done in December. I don't know when I'm actually going to have it done. I'm still working on it. It's really, really close to passing most of my tests. I practice test-driven development. I added a couple of (what I think are) really cool visual features recently. In the early morning hours, I work on SSIS Catalog Compare or training. I'll do that up until around seven o'clock or so. I'll make some breakfast for me and the family.

I'll read some Christian material while I'm having breakfast and spend a little quiet time. Around eight o'clock, I usually have at least one consulting gig going. I'll remotely log in to work. I work with Kent Bradshaw, who is currently coauthoring a book with me tentatively titled "Frameworks." Shannon Lowder is actually coauthoring the book with Kent and I. Shannon is another BimlHero. We're hoping to have that book out in mid-2018 with Apress— another Apress book. Kent and I work together a lot. We pair program really well. Kent can do data integration with SSIS. He's got decades of experience as DBA, but he can also do data integration.

I come at data integration from a software developer's background. Together, it's really like we're more than just two guys working on it. Pair programming is just an awesome way to do this. When we hit a project, especially if it's data integration with SSIS, we pair program. We also do a lot of performance-tuning work, a lot more than I thought we would ever do. We've been doing a lot of SQL Server performance-tuning lately. As long as we keep delivering results, clients keep calling us and paying us.

Mala: Awesome. Describe how being a professional can expand skills and embrace the new world of analytics.

Andy: Wow, that's a great question! If you're doing data integration, BI, that kind of stuff, these days, it's not a big leap into what I consider analytics. I would say analytics has always been the end goal of any kind of data warehouse or business intelligence solution. It's always about the reports.

You want to present information in an actionable format. You want people to be able to glance at a report and make a business decision.

Mala: Right.

Andy: You also want that data to show up really, really quickly. You don't want people to have click and wait watching a spinning wheel.

Even now when I train ETL people, I tell them, "We're doing the work to make these other people look really, really good—if we do our jobs correctly."

We format the data. We cleanse it. We educate our clients about what we've built and how it works. Then the report writers and the analytics people can then pick it up and go from there.

I got into my first data warehouse gig back in 2004 from the reporting end. That's where I started. I came at it from that background, but at the same time, I'm happy to admit I don't have a really good eye. Presenting the data visually in a way that's compelling, that people can automatically see what's there, that's just not my strength. Realizing that motivated me to surround myself with people who have that visualization strength. I can write a report. A couple people have shared with me that when they look at my reporting dashboards, it "looks like an engineer built it." Which, is true. I'm an engineer. The first time I heard that, I thought it was a compliment, but it turns out it's not. Being around people and having folks around who have that eye for both graphics and communication, it's very helpful for me to do that.

Making a transition into analytics these days though is a lot easier. It's even easier for people like me who don't have a good eye because the graphics are just beautiful in these tools. I mean, if you look at tools like Power BI and R. The graphics built into these platforms, they're just beautiful, and it's hard to mess that up. The tools are really easy to learn compared to their predecessors. I love Power BI. I love dashboarding. I'm drawn to that because that's what I was drawn to do back in my early days in manufacturing. I really wanted to produce these charts that would allow you to drill down. I would start with things like an AutoCAD drawing of the plant.

I built a manufacturing execution system [MES]. I didn't know what it was called at the time I built it. It's sad that I didn't know what it was called, but it was a manufacturing execution system. It was completely web-based. This was in the mid-nineties. I called it Plant-Wide Webs [PWW].

I'd take a drawing of the facility, turn it into a GIF—I'm never sure how to pronounce that word. You could turn that into a map in HTML. I would present that first. It would be colored using stop light colors: It would be green, red, or yellow based on the parameters that you'd set. The plants doing well. There's trouble. Or you're in big trouble. Those were the three colors that I used. It was easy to do that. Well, if you saw that you're in yellow, the plant engineer, or the maintenance people, or the business people for that matter, could click on the GIF and that would open up to a bunch of mini-GIFs that would also be green, red, or yellow. So you could continue to drill down in this, and of course, it was all manual.

I see a lot of that same concept in things like Power BI, where you have graphs and charts and tables, and they're all interconnected. I love that about Power BI. You just drag data onto the surface. You drag a few fields in. You do a little bit of configuration, then in five minutes, you can do what it took me five months to do back in the old days.

What I would share with folks is that gaining the knowledge is a first priority. Understanding the tool sets and being a more valuable person, either as a consultant to your clients or as an employee to your employer, either of those is good.

Having certifications is probably one of the best things I've done in my career. I took a few certification training courses near the beginning of 2017, but I really focused on it the last quarter of 2017. What I noticed was, one of the metrics that LinkedIn gives you is how many people have viewed your profile in the last ninety days.

My number was hanging out around one hundred fifty. It may go up and touch three hundred, but that's rare. Around two hundred was normal for me. I started completing these certifications and adding them to my profile. I've added three in the past four months. My LinkedIn profile views are now up over five hundred in the past ninety days.

Mahadevan: That's extremely valuable. Wow.

Andy: I have a heart for DBAs. Mala, I'm not a DBA. I've never been. I tried to do the job. I've almost been fired. In fact, I quit before I was going to get fired. I am that bad at it. I appreciate good DBAs, and I understand the job and the role. I love DBAs. I don't think it's going away. What concerns me about the DBA profession is that I do think that there's going to be, at a minimum, a plateau. I suspect there's going to be a dip in the numbers of DBA jobs available in the future. Not because there's less work to do, but because there's more automation to accomplish that work.

That said, knowing what I know about DBAs and the profession, and the intelligence and integrity that it takes to be a good DBA, it's really more about attitude than anything else. They can do it. They can make the leap. They can go straight to BI. They could go to analytics. There's no doubt in my mind. If they want to. Like I said, it has more to do with my desire and attitude than anything else. Attitude and preference. Some people don't like to code models. And I get it. I do. I love it. There's definitely a path there for folks to make that transition.

I'll do another shameless plug.

I'm cohost of Data Driven, which is a data science, artificial intelligence, and machine learning podcast. People can check that out at datadriven.tv. We're at one hundred episodes right now. There are two episodes we've done so far. Two really focused on this. One was with Buck Woody and another with Darren Lacey, both from Microsoft. Really good shows. Check them out for more information.

Mala: What's your experience with agile and business intelligence?

Andy: That's a good question. Agile and BI go well together. I'd also say that agile on steroids is DevOps. I think those go well together as well. I've found that agile works well with BI. I think the reason is that agile makes a couple assumptions. One is that you do not know all of the requirements when you start. That's particularly true of BI projects. In fact, I've never done a BI project where we had complete requirements when we started the project.

I've done some BI projects where we thought the requirements were complete, but it turns out that they weren't. Doing agile—and I like Scrum as an Agile methodology—I've had a lot of success with Scrum. I've also seen it fail. When Scrum fails—I've not seen it fail because it was applied incorrectly—I've seen it fail when it's applied to a problem that Scrum really wasn't the best solution.

Mala: Correct.

Andy: I think it involves wisdom and probably experience. To say Scrum's going to serve you, there are a lot of things that go into that. Probably the biggest factor is the culture of the team. Their experience counts. The way they solve problems. Being able to walk into a team and kind of detect how they're going to respond to stuff. That's just experience as well. My wife has a wonderful saying about experience. I'll share this, then I'll tell you why I'm sharing it. She says, "Good judgment comes from experience. Experience comes from bad judgment." I share that because I still make mistakes. I still exercise bad judgment.

Mala: Describe your experience with cloud adoption.

Andy: I've had some experience with AWS [Amazon Web Services]. Most of my experience is with Azure. The courses I've been taking have been designed around HDInsight and a lot of the open source stuff. Apache Storm, Spark, Hadoop—technologies such as these on the Azure. I've got a little bit of experience with ADF—Azure Data Factory. I'm focusing lately on ADF version 2. I actually got a call from a good friend last night asking about my experience with ADF, and I had to fess up and say I'm not yet billable. But that is something that happens in 2018. I will reach a confidence level with it, where I feel I am billable with ADF.

I've got quite a bit of experience working with databases and the cloud, and both Amazon and Azure. I've been doing some work with a MySQL pattern for about ten years now.

Mala: Oh, my.

Andy: I didn't feel comfortable sharing this MySQL with others because I'm kind of a Microsoft person. I was an MVP for five years. The design pattern uses SSIS to load MySQL.

MySQL didn't have a Microsoft connection. Gosh, I think it wasn't long ago—maybe May or September of 2017—that they announced MySQL in Azure. As soon as the announcement came out, I actually added this pattern to a presentation I'd been delivering on performance-tuning data flows in SSIS.

I run a data flow in SSIS that loads maybe twenty to thirty rows per second from a traditional SSIS data flow.

In the pattern, I replace the destination adapter with Script component. I use .NET ODBC objects and .NET code. I'm able to get that thing to run anywhere from a hundred to a thousand times faster. It's a sweet little pattern! I wore out three sets of elbows patting myself on the back when I figured it out about ten years ago. I took a nine-hour load down to twenty minutes.

Mala: Oh, nice.

Andy: Yeah. It's a sweet little pattern, but now of course AWS has Aurora, which is a rewritten MySQL implementation, and what I found is the exact same load pattern when I load it. It's still challenging when you load Aurora. It's not MySQL or Aurora's fault. It's the way that the SSIS destinations work.

The destination adapters in SSIS—and I promise I'm not criticizing them, I'm just saying that you've got to realize SSIS is built to be a very generic provider-driven, provider-agnostic solution. While that makes for a very broad spectrum of things that you can do with SSIS, sometimes you don't get the best performance out of that broad spectrum. There are certain places where you do.

I was sharing with someone earlier today that I probably have about twenty thousand hours doing SSIS development at this point. Over that time, you kind of develop a feel for what will perform well and what will not. People will ask me to articulate why something is a good idea or why it's a bad idea, and I don't remember why. I'm not very good at debating. I just remember that "this" is a better idea. That's how my mind works, unfortunately.

Mala: What are some of the new technologies in the big data and BI world that you are following and that you recommend SQL Server folks should keep up with?

Andy: SQL Server itself, and the velocity of change and evolution of the product. I find SQL Server very impressive.

The code the Microsoft SQL Server team has been putting out—keeping up with it will keep you busy!

Mala: I'm sure, yes.

Andy: There's that. I focus on the data integration in SSIS part of SQL Server, which is where I do most of my work. I'd say it's worth your time to learn at least a little about Spark, a little about Storm, and a little about some of the other engines that are out there. If for no other reason than to get a broader sense of what's available in the data integration market.

You know me, Mala. For years, I've been trying to build tools and utilities and patterns to facilitate test-driven development gradually in SSIS. I've built an SSIS execution framework that I've been giving away for years now. I mentioned Catalog Browser, another free tool that surfaces everything in an SSIS Catalog in a unified view. The SSIS Framework Community Edition that I give away is not only free but open source. I've got another collection of reports, Catalog Reports—also free and open source—that surface SSIS Catalog data using SQL Server Reporting Services [SSRS]. The SSIS Catalog reports that ship with SSMS are built into SSMS.

I've got a template in Power BI that's a very simple basic SSIS Catalog report. There are some much better fantastic Power BI dashboards out there for the SSIS Catalog, but this one will get you started.

Mala: What do you think about some of these smart data cleansing tools that are coming out? SSIS used to do a lot of that kind of stuff before.

Andy: Yeah, there are a lot of tools out there.

There's a little bit of blending going on from a Microsoft perspective in Azure with these open source Apache tools showing up in Azure HDInsight. One of the things that I found fascinating and a little validating is I built design patterns and mechanisms like the framework. I've written about how to overcome SSIS limitations that, telling folks, "You can do that in SSIS, but it isn't baked in. You have to write extra code to make it do it." It's possible in SSIS if you hand-code it, but in other engines it's just there.

To be fair, SSIS is technically free. Some of these other engines can cost you a million a year.

What I found validating about studying the other engines, especially when I was working with Apache Storm, is some of the design patterns I documented are already built in to Storm. They built it to scale out from the ground up. So they built it to run in parallel and to share workloads across servers. Hadoop is obviously built that way. When you start to put these technologies together, what I found validating is that they were solving the same problems I was trying to solve with SSIS. Things that were missing or needed to be manually built in SSIS. Things I was trying to add either with instrumentation or with frameworks or utilities, trying to make it easier, faster, and more flexible. Some of those things are just there in Storm. That's the sort of thing that you get when you look at another language and other platforms.

Mala: What's the role of documentation and communication in what you do?

Andy: Communication is actually the most important skill that any professional can have—period. Not just data professionals but any professional. It's more important than being technical. I say that and wince a little because I have failed to communicate well at times. It's not a strength. For somebody who speaks and writes, it's a little bit of an occupational hazard to not be really good at communication. I'll say I'm a stronger editor. I enjoy fantastic opportunities, and I'm surrounded by people who are very good at editing and cleaning up weaknesses in what I write and what I say, and providing great feedback. The community is actually pretty good at that when I give talks at SQLSaturday at the PASS Summit.

I take most of that feedback seriously. The only thing I look at and shake my head these days is when somebody complains about the level of a presentation.

Mala: What are your favorite books, blogs, and other means of learning? I know you mentioned certifications before, so other than that.

Andy: I read a lot. I balance my reading between fiction and nonfiction. I almost approach it like blending dishes in a good meal. I'm usually reading anywhere from two to five books at a time. I mentioned spiritual stuff. I read a lot of Christian literature. I reread books that I found helped me in the past. I'm rereading one now: Mere Christianity by C. S. Lewis. That's a good book. Also a lot of technical books. I read a lot of business books. I'm partial these days to audiobooks. I like Grant Cardone.

A few years ago, I listened to the audiobook, The 10X Rule [Grant Cardone, (Wiley, 2011)], and it appealed to me because of confirmation bias. The premise is that it's going to take you ten times as much work as you think to accomplish something. It's going to be ten times harder than you think, and you're just going to have to put your head down and bull through it. You're just going to have to outwork the problems. As a person with a work ethic that drives me to get up at five a.m. and work till somewhere between five and eight p.m., usually I identify with that. I think there's no substitute for hard work.

I'm a person whose background includes some time spent on welfare when I was younger. I have been on food stamps. I've made two dollars and fifty cents an hour shoveling manure out of stockyard stalls. That was a job I did when I was younger. Hard work is just something I think is essential. I think you get so much more out of that than just an appreciation for hard work. I think you get something that you can't get any other way. Self-respect. Dignity. That feeling you get when you accomplish something. I can certainly give somebody food. I can give them money. I can give them shelter. I can't give them that sense of accomplishment.

Mala: What are your recommended ways of stress management and having a healthy work/life balance? How do you handle long grinds on the job, or things that are boring and not challenging, and stay motivated?

Andy: Well, I'll say this. There was a time when I agreed more with people talking about passion and "do what you love." I even wrote about "do what you are" in a book on management called Managing Geeks. That was kind of an "in" thing at the time. That's since fallen out of favor. People are saying different things about it nowadays. They're saying, "No, don't do your passion." Grant Cardone, I think, has an interesting balance on this sentiment in that he says, "Become passionate about what you do." I now believe that's a better approach rather than to tell people to do what you're passionate about. That said, I happen to be very passionate about what I do. For me, it's not a choice. The passion was there already. I think I was born a geek. I really enjoy what I do.

When I talk about putting in long hours, it doesn't feel like long hours. So on work/life balance? I'm probably not qualified to speak to that, or at least not well qualified. I don't do a very good job with work/life balance. I am blessed with a wife who is very tolerant of my driven work ethic. She understands me. Even when she doesn't like it, she understands. She's very patient. Almost all the time. Sometimes she's not, but it's rare. I guess the best way to describe where I am now is that I'm trying to intentionally spend more time with my family and do things apart from work, especially as I get older.

Stevie Ray is our oldest child and he's turning fifteen in a few weeks. If you do the math on the number of weekends between now and when he turns eighteen, there are about 150 of those weekends left. It's like, do I want to work all of those 150 weekends? Do I want to spend all of those 150 weekends split between work and presenting at events like SQLSaturday?

I have a passion for sharing, and I love speaking at SQLSaturday. I take Stevie Ray with me to a lot of SQLSaturdays, in fact. He's been going to conferences with me for years. The reason I do that is it gives us time on the road to hang out. I don't do a good job at work/life balance to be honest with you. Working from home helps some because I can stop work, most days, around five o'clock or six o'clock p.m. Most days, that's the time to push the button and walk downstairs from the home office. But yes, I stink at work/life balance.

Mala: Describe your style of interview of a data professional? What do you look for in someone you interview? What questions do you ask?

Andy: I like to hear the phrase "I don't know." I go for that. It's subtle. Of course, I'm trying to determine where somebody is and ascertain where they are on a scale, especially if I'm interviewing someone for an SSIS position. I led a team when I worked at Unisys. We grew to forty ETL developers. I wasn't managing all forty developers. I was managing six or seven, and they were managing the rest.

I interviewed several of those folks. We got into a spot where it was just hard to find people. I joked and said, "If you could spell SSIS, I would hire you." The truth is I was looking for people who had deployed an SSIS package and had it run. That was my minimum requirement.

Even then, I had a couple people I interviewed that hadn't done that. I hired one of them. I've been training people for over thirty years. I did training back before I knew anything about databases. I was delivered electronics training with the military. I taught at a technical college here in Richmond, Virginia, called ECPI. I taught electronics and math. I think I taught one semester of DOS, even. I knew I could train these folks, Mala. I knew I could train them, and I did. What I'm after more than the technical is, "Are you going to be straight with me?" If I ask you a question and you don't know the answer to it, I never need for you to try and BS me. I need for you to tell me you don't know. There are things that I don't know. I learn stuff every day, every gig. I tell customers going in, "You know stuff that I don't know. I'm going to learn stuff here."

When I interview people, I want to hear them say, "I don't know." If I can't get an "I don't know" out of them, it's not the same exact thing as lying to me, but it's pretty close. One of my interview questions on SSIS is, "Which is better, to use a Data Flow task or an Execute SQL task?" And the answer to that question is neither. Or, "It depends."

Mala: It depends. Yeah.

Andy: That's the actual answer. But what I want to hear that. Then I want to follow up with, "Tell me why you think that."

I want to hear the rationale. I've had people answer that question, "Execute SQL task," and give me an answer to why they think that. I've had people answer, "Data Flow task," and give me a good answer as to why they think that. I've hired both. What I'm more interested in is if you understand there's a difference, and your rationale for your answer.

One of the people I hired answered that they would use an Execute SQL task because they really didn't understand enough about the Data Flow task to feel comfortable building data integration solutions in an SSIS Data Flow task. I hired that person because they were honest with me.

Mala: Right.

Andy: I spent two days training them, and in about fourteen hours, they knew what a Data Flow task was. They knew how to use it. I just taught them what they didn't know. But I would have never done that if they'd tried to put one over on me and tell me a Data Flow task wasn't the right task to use. I would never hire a person who decided that it was better to try and bamboozle me than just tell me, "I don't know."

I've done the same thing in this interview. I've shared with you that I'm not a DBA. Can I do performance tuning? Yeah. Am I the best at that? Heck no. I call other people when it's time to get down and dirty serious about performance tuning on the DBA side. I can performance tune SSIS.

That's really what I look for. I'm trying to find integrity.

Mala: Why do you recommend that people be involved with the community? What are some of the suggestions you have about that?

Andy: I used to say that I was self-taught. I wasn't born with this knowledge. I'm not that smart. I'm not a particularly fast learner. I'm a thorough learner and that makes me slower. I'm a little dyslexic and a little ADHD, so those work against me. The SQL Server community is the best community—I think—on the planet. It goes beyond technical. People actually care about people. It shows.

There's no harboring or hoarding of information. Steve Jones and Andy Warren have written extensively about networking at events, especially the PASS Summit. I couldn't agree more. I shared that experience with you earlier about Ken Henderson. It literally changed my career. It's not just the knowledge I pick up, although I pick up a bunch when I attend events. It's the networking. It's getting a pulse on what's going on in the technical community at large by talking to people I know as well as always making new friends. I do that every time I attend a SQLSaturday, and every time I attend a user group meeting in Richmond, which is about seventy-five miles up the road. Whenever I attend something like a PASS Summit, I always meet new people. I make new friends. Social media is another great way to keep in touch when we can't see each other.

Mala: Are there any specific people you follow for technical reasons on social media?

Andy: Absolutely. Brent Ozar. Cathrine Wilhelmsen, who blogs a lot about Biml. Kendra Little, Paul Randal, Kimberly Tripp, Chris Webb, Brian Kelley, and Steve Jones. I'm trying to think of some integration services folks that I follow and read. There's just a lot. There's this whole community of SSIS people. You know a lot of people who probably consider themselves new and newer. I like to read those early entries, and I try to encourage them, leave a comment. I remember the Datachix, Julie Smith and Audrey Hammonds. I've learned stuff from both of them. Rafael Salas, who has kind of moved away from SSIS. Matt Masson's old blogs are still gold!

Mala: Last question. If you had one superpower, what would it be and why?

Andy: Oh, I don't trust me with a superpower. I know me, and I shouldn't be trusted with anything like that! I do a bad enough job with what I've been given already.

I'm thinking through comic book characters—my heroes when I was a kid. I was a Captain America fan. I used to read Captain America comic books. Captain America was very strong. They represented that well in the Marvel movies. Very strong physically and as an individual. I guess that could be it. I could be stronger—physically stronger.

What I really admire about Captain America are his morals and his integrity. I think that's important. You should not be afraid to stand up and speak up when you see something that needs addressing.

Mala: Of course.

Andy: I've done that, and it's cost me. I'm okay with that. I'm more than willing to pay the price I paid for standing up against some of the stuff I've stood up against. But a superpower? Yeah, physical strength would probably be nice because I don't feel very strong. What would I do with it? Gosh, I don't know. Pull up trees out back here. I need to knock down some trees, and I could just pull them up out of the ground by the roots. That'd be kind of awesome.

Key Takeaways

- Failure and imposter syndrome are normal. You are better than you think you are. You know more than you think.

- Present information in an actionable format. You want people to be able to glance at it and make a business decision.

- Certifications boost LinkedIn profile views considerably.

- There's going to be, at a minimum, a plateau, and probably a dip in the numbers of DBA jobs available in the future. Not because there's less work to do, but because there's more automation.

- Good judgment comes from experience. Experience comes from bad judgment.

Favorite tools: Biml, SSIS, Power BI

Recommended training: edX certification courses

Blogs to follow: Brent Ozar's blog, Kendra Little's SQL Workbooks blog, The Datachix Blog, SQLskills.com, Cathrine Wilhelmsen's blog

Jes Borland

Premier Field Engineer, Microsoft Corporation

Jes Borland is a premier field engineer at Microsoft. She has experience with a wide range of SQL Server features (versions 2005 through 2017), and is on the cutting edge of Azure technologies. Since graduating Fox Valley Technical College with an IT-programmer/analyst degree, Jes has worked as an SSRS developer, DBA, and consultant. Her involvement with a variety of implementations has created a deep well of experience from which to draw as she works with customers. Before joining Microsoft, Jes was a six-time Data Platform MVP, recognized for her community work. Jes tackles every project and problem with tenacity, and her enthusiasm is unmatched in the SQL Server community. In her free time, Jes never stops moving—counting fitness, coffee, cooking, and travel as essentials to life. You can find Jes on Twitter (@grrl_geek) and LinkedIn (www.linkedin.com/in/jesborland), and learn from her SQL Server challenges via her Less Than Dot blog (http://blogs.lessthandot.com/index.php/author/grrlgeek/).

Mala Mahadevan: Describe your journey into the data profession.

Jes Borland: My first technology job was at an Internet service provider doing customer service—just answering the phones about paying bills and some really easy tech support questions. After I'd been there about a year, they had taught me everything I needed to know about dial-up Internet and

© Malathi Mahadevan 2018
M. Mahadevan, *Data Professionals at Work*,
https://doi.org/10.1007/978-1-4842-3967-4_5

modems, and I started writing scripts to help with mailbox clean up. I decided I wanted to go to school for technology. I initially had the idea to go into web design.

I was going to school part-time and working full-time. I eventually transitioned to a different company doing more help desk work. In school, I took a Cisco networking class and a SQL class—structured query language—writing against the mainframe. I remember struggling with the networking class, which was one of the required classes, and I decided I definitely did not want to do system administration or networking stuff permanently. But I loved the SQL class. Loved it! I finished the required work in nine or ten weeks, and then I came back to finish the semester to get harder exercises from the teacher and to help other students.

I worked at a company of about sixty people. Our IT department was about eight people. Our database administrator had a backlog of report requests and he found out I'd taken an SQL class. He helped me install Reporting Services 2005 on the computer. He handed me a stack of report requests and said, "Have fun!" I did that for about a year, so I eventually transitioned into what we called the "decision support specialist" at the time—before "business intelligence" was a buzzword. I loved it, and that's when I started blogging as well. That's when I started giving presentations at user groups in SQLSaturdays.

Eventually, the DBA left the company, and I took on that role. Another person in the company—he was actually from the finance department—transitioned into the report writer, so I got to teach him a little bit, which I thoroughly enjoyed. Then it went from there. I progressed to being a DBA at a Fortune 500 company. I went from a very small company to a very large company with a lot more servers and a team of DBAs. There was automation, PowerShell, performance tuning—things that I didn't have to worry about previously. Then from there, I went on to different consulting jobs, teaching, and now working at Microsoft.

Mala: Describe a few things you wish you knew when you started your career that you know now and you'd recommend to people who are new.

Jes: There are two things and they both have to do with learning. Number one is no one person knows everything. At my first DBA job, I spent a lot of time feeling intimidated by the DBA. It seemed like he knew the answer to every question I asked. I assumed he knew everything, and I felt very incompetent next to him. I failed to forget that at some point in time he had started where I started. He wasn't born with this vast knowledge of SQL Server and databases. He had to learn.

Mala: Right.

Jes: The same thing happened when I transitioned into the other big company and even now as I continue to do consulting or work with people in different enterprise companies and different industries, no one person knows everything. Don't be afraid to ask questions. If someone gets to the point where they just continue to answer all of your questions and they never admit, "I don't know," and they never say, "I need to talk to someone else," don't trust them. They are making stuff up at that point.

The other thing I would say is, with the amount of data that is being produced in the world right now—the amount of data being retained and how businesses are trying to make decisions based on the existing data, we have to continue learning. Being in the data field is going to entail lifelong learning.

There is no point where you get a degree, a certification, or a specific job, and then get to stop. That is not our world. I would say that for people that are passionate about lifelong learning and continued growth, technology in general but specifically the data field are exceptional fields to go into.

Mala: That said, do you have thoughts on how to keep up? Or is it too much and you're not sure you're learning enough?

Jes: My usual struggle is what I should be learning. I'm very experienced with the internal database engine, performance tuning, and high availability. Should I be spending time learning about big data? Should I be spending time working on machine learning or statistics? Should I learn Power BI? Should I learn more about the cloud?

What I try to do every year is establish one thing that I want to know more about. It can be related to something I've done. It can be brand-new. But it's something I'm going to learn. What I want to do in 2018, for example, is work more on big data, like data lakes, Hadoop, all of that stuff because it's related to SQL Server and database engines. But it's different enough that I can learn something new.

How to fit in the time is always a question I get. People say, "Well, I have a job, and I have a family. I have a life outside of work." My answer to that is, like anything else in life, if it's important to you, you will make time. I put time into my schedule. It's on my work calendar and it's on my personal calendar. As you know, I am also a big runner. Running is important to me, so I block out time for running and for yoga every week, the same way I block out time for learning every month.

Do we make every single session? No, and we never will. We're not perfect. But again, if it's important, you'll make time for it.

Mala: What's a typical day in your life as a professional?

Jes: That's a good question. My current role consists of two major roles. I provide support for my customers, and I provide training for my customers. In a typical week, customers may have a question about something with SQL Server, whether they are doing an upgrade, experiencing a performance problem, or running into a problem with a feature. They contact me, and then we work together to solve the problem. I write scripts. I provide them with resources to solve the problem. I help them troubleshoot, and I love doing that. I think troubleshooting is a lost art. I think that problem solving is something that we can all be better at. That's why I love doing it.

Another big part of my job is training and teaching, because SQL Server is such a huge platform now. There are so many versions and so many features. I regularly spend time writing articles on how to do things. I write training materials, whether it's a one-hour or a two-hour chalk talk, or a full day, or even a full week of training on a more in-depth topic, and then I present that information.

Mala: What are some industry trends that you're looking out for?

Jes: We have the cloud, we have big data, and we have privacy. I think using the cloud is becoming more and more important to companies of every size and to all of my customers. Cloud services provide new features or more scalability. Customers of all sizes are interested. The major cloud providers—Microsoft Azure, or Amazon Web Services, or another type of cloud service—are all investing heavily in new features, more regions, more data centers to make those offerings more attractive and lower priced for customers.

Companies that said, "I'll never be in the cloud," are coming to the cloud. As a data professional, it's important to understand that the cloud isn't replacing on-premises for most companies, it is complementing it. It's enhancing it. I think that's important, and so every data professional continuously learning about cloud services and what's available is really important. Hand in hand with that, of course, both on-premises and in the cloud, is security. All the time, we hear on the news about data breaches and information being stolen. But we only hear about a fraction of what actually happens.

As data professionals, we deal with sensitive information, and we need to be constantly learning about, thinking about, and aware of how we are protecting the data that we're managing. I think every data professional needs to be concerned with that, whether you are the database administrator who is making sure only the right people have access to these servers, to people analyzing and writing reports. Think about things like, "Do the people I'm writing this report for really need to see this data? Am I making sure that only the right people in the right departments have access to it when it's provided?" That's end-to-end life cycle.

Last but not least, I would say another major industry trend is, to use the buzzword—big data.

Mala: Right.

Jes: As someone who has worked almost exclusively with SQL Server over the past ten years, my world is very relational data but every type of company collects multiple types of data now.

There are some things that fit nicely into relational databases but there's also a lot of other data that does not—unstructured data, images, videos, documents, sounds, sensor data. Companies are collecting and want to analyze and use that data for reporting and forecasting, and figuring out what's coming next. As a data professional, we don't need to know how to manage every data platform. We don't need to know every language. We don't need to know SQL, U-SQL, PolyBase, and six other languages.

What we need to understand is that when the business comes to us with a request to store data and access it, we really have to think critically. What kind of data is it, and are we using the right and best method for storing it and retrieving it? This means that, at least at a high level, we have to understand these other technologies and how they work.

Especially in small to medium-sized companies, there may be one or two people that deal with data. Larger companies typically have departments for each of these things. It helps if you're a relational data expert. If you're working with someone on big data or unstructured data, it helps when you can at least speak their language, know the terms, and know what's important to them.

I think those are probably the three big trends that I see—using the cloud, securing our data, and expanding beyond relational into big data.

Mala: What are some of your favorite tools and techniques that you use on a daily basis with DBA work?

Jes: For database administration, to me the single most important thing is automation.

When I started as a DBA for SQL Server using Management Studio, I learned by using the GUI. Everything is click here, right-click here, go through this wizard, fill out this screen. I soon realized that if I was trying to design a view or run a backup, and I went through the GUI every single time, it took a lot of time. So I learned how to script things out.

I always say, if I do something once, save the script. If I do something twice, it's time to automate it because chances are it will need to be done again.

Learning T-SQL scripting is not just for report writers or developers, it's for database administrators as well. They need to learn that fundamental language so that they can automate as much as possible. If you want a tool for automation, it's PowerShell—especially in the Microsoft ecosystem.

I have a great story about PowerShell. When I was a DBA at the Fortune 500 company, one of our tasks was to resolve failed job tickets. It could be a failed backup, a failed restore, a failed index maintenance—something like that. We logged in to our ticketing system to get the server name and the job name. Then we would have to RDP [Remote Desktop Protocol] into the server, enter our credentials, open File Explorer, go to the directory where the job log was written, read it, and then figure out what was wrong and resolve it. The job logs were written to the server, not a central repository.

It would take a few minutes per ticket to just find out what went wrong. We could easily have eighty-plus tickets a day, so it really took a lot of time.

I took my first PowerShell class and learned that you could remote to another server, navigate down into a directory, and read a file, so I wrote a PowerShell script to do just that. You could open a ticket and then open your PowerShell script and enter a server name and a job name. It was a highly structured system at an enterprise company where every server had the same structure for the log files. Entering it would bring up the job log right then and there on your computer, no RDP, and then you could go into Management Studio and restart the job, or go to the backup system and restart the backup if necessary, and then close the ticket. It went from up to five minutes to solve a ticket to between thirty and ninety seconds to resolve a ticket.

Mala: Oh my!

Jes: Take that eighty times or more a day and we calculated that we were saving something in the area of $25,000 a year by doing that. That's when I learned a lot of PowerShell. Automating with PowerShell is fantastic.

I also do a lot of work with Azure—using the tools that are available. Azure Automation has a lot of powerful features for making things happen in Azure with not only virtual machines but also things like Azure SQL Database, Azure SQL Data Warehouse, and other built-in tools.

Other things that I use frequently… In terms of database administration, the community has a lot of fantastic scripts and tools that they have built—things that you don't even necessarily have to pay for. Tools that I love using include Adam Machanic's sp_whoisactive. That's a fantastic tool for troubleshooting.

I also really enjoy Ola Hallengren's SQL Server Maintenance Solution. I find a lot of companies use that because it just makes backups, restores, indexing, and statics. Very easy.

In general, I think that if you are looking for a tool or a script to do something, the SQL community has it on GitHub.

BPCheck.exe, or the Best Practices Checker, is another really good tool from the Microsoft SQL Server Tiger Team. It's a T-SQL script you run against an instance. It checks to see if things are set up according to what Microsoft thinks the best practices are for SQL Server instances and gives you a list of things to work on if they're not. Those are some of my favorite tools that I've been the last couple of years.

Mala: What struggles have you faced with management or the business side of things? What's your recommended approach to handling it?

Jes: One of the biggest struggles that I have had with businesses over the years in terms of technology and data is getting companies to invest in regularly updating to new versions of software before things go out of support.

There are a lot of companies that have a "if it's not broken, don't fix it" mentality. I don't think that that is always useful.

Mala: I agree.

Jes: It's always that, or "we heard that the licensing model changed, we don't want to end up paying more money," or "this system is so critical we can't possibly take an outage of even a couple of hours of downtime to upgrade it on to something that's newer."

How do you convince the company it is in their best interests to upgrade to a newer version? The best thing that you can do is to set up a development environment with the new version and move your data over. We're data professionals—we provide data about how things are better. Perhaps a new compression feature has been introduced that could reduce size on disk by X percentage, which means X amount of dollars and savings. Perhaps some new features have been introduced into the database engine that makes things faster, so your top five queries that were taking a long time now run twenty-five or fifty percent faster, meaning that the business will get its data faster. That can be a really powerful tool.

Besides just saving money, one of the big industry things right now is security. It's making the company aware that older versions, particularly on database software, may not have all the security features that newer versions do. I know with SQL Server 2016, for example, when the row level security and Always Encrypted features became available, there was a big push for companies to say, "Hey, let's start using this newer version of SQL that allows us to keep our data more secure and to fit in with privacy laws and privacy regulations."

Mala: Right.

Jes: As data professionals, we shouldn't say, "Hey, this is the latest and the greatest. I want to play with this." It's knowing what the businesses' pain points are. Know what the biggest applications that have the most sensitive information in them are. Know the applications that are the oldest, or the slowest, or that require the most performance. When you see that there are things that can be taken advantage of, speak to the business. Let them know that we can save this amount of money, or we can save this much time.

Don't talk about things like upgrades in terms of "this is cool, or exciting, or shiny and new." Instead, say, "This is how we're going to solve a business problem."

Mala: What sorts of issues are caused when you interact with other technologists, like developers, SAN admins, VM admins, and so forth? Are you likely to clash with them, and why? What do you recommend for talking to people in other areas?

Jes: A lot of people don't take the time to learn the language of other systems. Much like I said, someone that's worked solely with relational data may not be able to understand how a nonrelational database works, or how a data lake works. You should strive to understand that or strive to understand big data a little. We have to understand how IT works in general.

I find that working with other teams is a struggle simply because we don't speak the same language.

I strive to learn a lit bit of their language so I communicate better with them. It's interesting that the relationship that I find is usually the most contentious is the relationship between application or web developers in database professionals, and it's because things are very different. Developers use very object-oriented languages.

They use languages where they tell the program exactly what to do and in what order, and that's what it does, whereas SQL is what we call the "declarative language." Given an idea of what data we wanted to bring back, we let the engine decide how best to do it.

Having taken many development classes in college, I understand that those things are very different. I also understand the developers are typically smart people that learn things very well, so rather than just saying they're doing something wrong or writing something poorly, it's a matter of educating them. Get them to understand one small thing at a time. Did you know that when you write a query like "select these columns from this table where this happens," that SQL Server doesn't process it in that order? We start with the from, and then we work on the where to narrow things, and then we do the select, so let's write our queries a little bit better to follow along with that. They're amazed at that indexing and how properly indexing for our queries makes a difference. Many developers don't understand the concept of indexing properly for SQL Server because no one's told them it's important. They just know that they need X amount of data back. Sometimes it doesn't happen fast enough, so they tell you the database is slow.

Mala: Right.

Jes: In every job that I worked, as a database administrator and as a consultant, when solving a problem, I always strive to educate the person I am working with on why SQL Server thought it was a problem. That applies to more than just the developers and things like how we write our queries and how we index for things. Getting some admins to understand the SQL Server data files versus log files and how it primarily writes to the logs files and reads from the data files, and how tempdb is used, and thus why that means it's a better disk, or a different disk, or its own disk.

Again, they may only be used for things like file systems, where the layout may not be as sensitive to performance impact as something like a database is. Again, when working with other departments or with other technologists, it's not just "there was a problem and I solved it," or "you have a problem and I fixed it." It's "there was a problem and here's why SQL Server thought it was a problem." Educating them just a little bit so that the next time it happens, they understand a little bit why and can fix it themselves.

Mala: What's the role that documentation and communication play in your job and how important do you think that is?

Jes: Communication is the number-one most important thing in anything. It doesn't matter if it's a working relationship or a personal relationship. Whether it is a business-to-business relationship or a business-to-consumer relationship, communication is the single most important tool we have. Clear communication—verbal and written—is incredibly important. Documentation is the second most important thing. People say they hate writing documentation. They don't want to take the time to document things because they'd rather be fixing another problem or making some cool new software.

I once had a teacher that would say things like, "The code tells you how to do it. The comments on your documentation tell you why you did it."

I always want to ensure that if I am ever away or someone else has to come and read my code, or administer my systems, or support the availability that I have built, they understand why I made the choices I did and what I did, which is why I stress that documentation is super important. Documentation is merely another form of communication.

Also, you can't remember everything—especially with how much technology there is, and how much things change, and how everyone and every company seems to be doing do more with less now. If you document things, you don't have to stress out about remembering everything. You can write it down and move on to another problem.

I have tons of documents: a SQL Server setup guide, a SQL Server post-installation checklist guide, a SQL Server uninstall guide, an Availability Group setup guide, an Availability Group troubleshooting guide, a performance tuning guide. I document a lot of things, and not only does it help me to not have to remember every little thing, it's something that I can go back and reference. They are also really good training tools and teaching tools because there are always new people entering the industry and new people that you're working with.

Imagine if when DBA starts, day one, you can give them a bunch of things to read rather than having to always just have them I'm looking over your shoulder. Let them know that asking questions is absolutely okay, but imagine just being able to read about a bunch of things. It's a lot easier that way.

Mala: I completely agree.

Jes: I think for me personally, blogging about technology has helped me hone my communication skills because I can even go back and re-read what I wrote five or seven years ago. Ooh, that was so bad and now I know, right? Once you write a blog and then you get enough people to leave comments on it, you realize, "I didn't answer this question. I totally skipped this step. I forgot to mention that little thing." That's the feedback that you get. You realize how you could have made it better, and then you start to incorporate things the next time you write something.

It's the same thing with speaking. You're presenting topics, whether it's for fifteen minutes or for a full day. Again, getting up in front of people to talk about a technology problem and solution really teaches you how to better communicate that information.

That's one of the reasons that I tell people to not just blog and speak simply to make a name for yourself. It's really going to fundamentally help your communication skills in business. I frequently tell people who are considering speaking at user groups or technical conferences, "Imagine this. You're on a team of DBAs and you have to come up with an upgrade plan. You and another DBA have different opinions on what your upgrade plan or new software plan should be. Management asks the two of you to give a presentation about what your plan is and why it's better. Who's going to have the upper hand there? A DBA who regularly speaks in front of other people about technological or even nontechnological topics and who is comfortable talking in front of people and handling questions, or someone who sits at their desk every day doing their job but doesn't talk to people?" They could have the better plan but nobody is going to know because they're not used to communicating. That's super important to me.

Mala: What are your favorite books, blogs, and other ways to learn?

Jes: Interestingly enough, I am a reader. I am not a fan of watching videos or recordings. I can't watch YouTube videos. I can't watch recordings from conferences. I get too easily distracted.

SQLServerCentral has a weekly newsletter. I make sure that I read that every week.

Mala: Database Weekly. That's my favorite. I love that.

Jes: Yes, the Database Weekly. That's my number one without a doubt. I also follow the SQLskills' blogs and the weekly newsletter. Again, they're very intelligent people that communicate well. They have a lot of interesting stories to tell.

I have a long list of bloggers that I read as well, and I find many of them on Twitter. I think Twitter is a great resource for following different hashtags and following different people. If it's a popular posting, something that's been retweeted two, three, or four times in the morning, I think, "Hmm, that must be a really exceptional one."

There is also an MVP named Alvin Ashcraft who has written a blog from our daily newsletter for many years called Morning Dew. It's not specifically a database administrator writing it, which is the fascinating part for me. I can look through there and get things about development, web development, or big data.

I also subscribe to the Power BI team's blog posts. Power BI is one of my favorite tools for visualizing data and doing a lot of things. I think that their team does an exceptional job of communicating and content. One of my favorite people who works with Power BI is Adam Saxton. He's the "Guy in a Cube" who does Power BI and Analysis Services videos, and things like that. I would say that those are probably my favorite mechanisms. I probably spend at least an hour a day reading articles.

The other benefit that I get out of reading blogs and books—rather than watching videos and recordings—is when someone shows how they solve their problem in a blog or a book, they have scripts and things like that that I can just copy and paste into a sandbox where I can play with it right away. I learn best by doing. That's why I think I choose reading over anything else.

Mala: Are there any nontechnical books that you really like that are useful for work?

Jes: I have one that I highly recommend to everyone. It's a book called Essentialism by Gregory McKeown [Crown Business, 2014]. Essentialism says that you cannot give equal weight to ten different things at once and expect to succeed at any of them. It's essentially "go big or go home." Throughout the book, the author gives an indication of how to do that. One of my favorite chapters is about how to say no. It is the number-one problem I think people have. Someone asks you to do something, whether it's at work or in your personal life. As human beings, we are conditioned to like saying yes. We like helping other people. It makes us feel good.

There's this whole chapter in this book about different strategies for saying no. I loved it, it's fantastic. My book is highlighted, dog-eared, and has notes written in it. It's probably the one book that I would tell everyone to read this year.

That chapter on saying no ... It sounds cheesy, Mala, but that chapter changed my life.

Mala: It didn't sound cheesy to me at all because I have a lot of problems with saying no.

Jes: There was one phrase in it in particular that I remember. "Saying no to an opportunity is not the same as saying no to the person. And if the person you say no to doesn't understand that, that relationship isn't worth having."

Mala: Very well said. Yeah.

Jes: Yes. If someone comes to me with something I can say, "Yes, I can do it but in a week," or "I can't do that but let me give you the name of someone who can." Those are just two of the strategies mentioned in that chapter. If that person gets insulted and angry, then that person clearly has some issues, and I don't want that kind of relationship. My gosh, that book and that chapter were just fantastic.

Mala: Describe your style of interviewing at DBA. What do you look for? What's your style of interviewing?

Jes: It is my firm belief that technical knowledge can be taught. Problem-solving skills and communication skills—the soft skills—are what I am looking for when I interview. I'm not going to ask a person to name trace flags, or tell me what the names of system views are, or obscure minutia that even I would have to look up online. What I want to do is present them with a scenario.

Here's an example. You have a three-node Availability Group. It's set up in such-and-such a way. The business wants to be able to expand it and do some reporting on it. How would you solve that problem? Or, give the interviewee a high availability and disaster recovery sort of problem to solve. We have a new application coming in that is super important. The business says that they can't lose any data and if there's a major outage, we need to back up in a second facility within an hour. What are some of the options that we have? How would you look? What questions would you ask them? How would you go about solving that problem?

One of my favorites to ask someone knocking on my door complaining that the server is slow. You don't have to tell me exactly what scripts you're going to use to solve that problem but what are some things you would check? What are the steps to take? I am always looking for problem-solving skills. How do you go about solving problems? The ability to clearly and thoroughly explain the steps that you would take is part of the communication skills that I am looking for.

Mala: What are your contributions to the community? Why do you recommend people be involved with the community?

Jes: No one can do everything on their own, and no one can know everything. Getting involved with your technical community is going to help you solve problems. It's also going to teach you a lot because when someone asks you a question and you can help solve it, you learn something as well.

By being involved in the SQL community through Twitter, blogging, writing books, or speaking, I end up learning more myself. Again, it comes back to that main topic of lifelong learning, and we are in this for the long haul.

It makes you feel good when someone sends you an email or comes up to you after a presentation and says, "That was amazing. This helped me solve a problem that I've had for six months," or, "This will be useful to me tomorrow." By sharing knowledge, you understand that nobody knows everything. You can help other people. I also think that the SQL community, in particular, is full of good people that are kind, helpful, and goodhearted, and it's not just when we have technical problems. I've seen people band together to help others that are ill, and to raise money for family members whose houses have been destroyed in fires. Again, you give, and people will give back to you as well.

Giving to the community and getting involved in the community has been one of the best things I've ever done. It has helped me make not only networking connections but also lifelong friends. That's amazing. It's not every industry that gets to say that these people that I talk to every day are friends. When I travel for work, no matter what city I go to, if I want to have dinner with someone, I can put a message out and without a doubt, I could find a friend to have dinner with. I love how the community comes together and helps each other.

Mala: If you had one superpower, what would it be and why?

Jes: If I had one superpower it would be teleportation so that I could travel anywhere in the world instantly. Then I could visit every place I want to go on Earth. I have a lot of vacations I need to take.

Mala: On the same lines, what are your strategies for work/life balance? What do you recommend? Because DBA work is typically stressful and people often struggle.

Jes: It is stressful—long hours, on-call weeks, deployment weekends, and you have to constantly learn. Number one, I block off time on my calendar specifically for certain things every week. I think that everyone should have some non-technology hobbies that they get involved in and make time for. Again, if it's important to you, you'll make time for it.

For me, it's always been important to set boundaries with my manager and my coworkers that if I'm not on call, I'm not going to be taking calls or checking emails outside of work. I stick to that because inevitably, if you slip once or twice and start doing that, people are going to expect that you will do it all the time.

Mala: So true.

Jes: It requires a little bit of discipline on your part, but you have to set those boundaries. It's another thing that's mentioned in the Essentialism book. Again, if you let the things slip once or twice, it will continue to slip and slide, and you'll be working every night and every weekend. Show a little bit of discipline. Set boundaries and discipline yourself to follow those boundaries.

The other thing that I tell technologists is to take a vacation at least once a year where you unplug completely.

Key Takeaways

- No one person knows everything. There are always things we don't know. Never be afraid to ask questions.

- If anything is important to you, you will make time for it.

- The cloud, big data, and security/privacy are the three big trends to watch for and get involved in.

- Don't talk about things like upgrades in terms of "this is cool, exciting, shiny, and new." Think instead of "this is what we'll use to solve a business problem."

- Give to the community and get involved in the community. It helps with networking and gets you lifelong friends.

- Talking about a technology problem and solution in front of people really teaches you how to better communicate.

- Take a vacation at least once a year where you unplug completely.

Favorite tools: Ola Hallengren's SQL Server Maintenance Solution, PowerShell, Adam Machanic's sp_whoisactive

Recommended Books: *Essentialism* by Gregory McKeown

Blogs/Newsletters to follow: *Database Weekly*, SQLskills.com newsletter, Alvin Ashcraft's Morning Dew, Power BI podcasts

Kevin Feasel

Engineering Manager, ChannelAdvisor Corp.

Kevin Feasel *is a Microsoft Data Platform MVP and the engineering manager of the Predictive Analytics team at ChannelAdvisor, where he specializes in T-SQL and R development, fighting with Kafka, and pulling rabbits out of hats on demand. He started his post-graduate career as a web developer working for the Ohio Department of Alcohol and Drug Addiction Services, where he was the slowest to run away from the database and thereby became the de facto database administrator. Later, he became the de jure database administrator for the agency. Eventually moving on from there, he did additional work as a database administrator focusing on data warehousing and ETL solutions. At the end of 2013, he became a database engineer at ChannelAdvisor. In January of 2017, he started his current role as the manager of a new team dedicated to predictive analytics.*

Kevin holds a Bachelor of Science in computer information systems and German from the University of Dayton and a Masters of Economic Policy from Albert-Ludwigs-Universität in Freiburg, Germany. He is the benevolent dictator of the Triangle Area SQL Server Users Group in Raleigh-Durham, North Carolina, and a co-organizer of

© Malathi Mahadevan 2018
M. Mahadevan, *Data Professionals at Work*,
https://doi.org/10.1007/978-1-4842-3967-4_6

the Triangle Area .NET User Group. He is a contributing author to the book Tribal SQL (Red Gate Books, 2013). He is also the curator for Curated SQL (www. curatedsql.com), which is a daily compendium of interesting blog posts in the data platform space.

A resident of Durham, North Carolina, Kevin can be found cycling the trails along the Triangle whenever the weather is nice enough. You can also find him online at www. catallaxyservices.com or @feaselkl on Twitter.

Mala Mahadevan: Describe your journey into the data profession.

Kevin Kevin: I had a background in computer science—that was my undergraduate at the University of Dayton in Ohio. I've always had a liking for stats and analytics, but I never really spent that much time with it outside of occasional coursework. At various jobs, I ended up doing things in Excel, but nothing serious.

After I graduated from Dayton, I got a DAAD Scholarship to study economics in Germany.

Mala: Oh my! Wow!

Kevin: Yeah, it was a lot of fun. I got to travel around and see a lot of Europe. I got a master's degree in something that was a topic of interest. I thought about getting a PhD, but I decided it was probably time to join the real world.

So, I landed at the State of Ohio as a web developer. I turned out to be the guy who was the slowest to say "not it" when there was a database problem, and that turned out to make me the DBA. A strange way of doing it, but they had their reasons.

It turned out that I loved that work. I really enjoyed picking up, learning a lot about SQL Server, and really digging into it more. As a result, I went from accidental DBA, to real DBA, to database developer, and then I went into analytics full time. I just kept doing those types of projects, and a position opened up at my current company running an analytics team. So, of course, I jumped on it.

Mala: Awesome. Fantastic. So, you're one of the DBAs who got there really early, looks like.

Kevin: Yeah, it was my first real-life job. So, is it the norm that people spend a lot of time elsewhere first?

Mala: Yeah, right now every single person that I've been talking to wants to get into analytics. Also, normally, companies tend to associate certain things, and one of the associations that I've been dealing with personally is that they think everybody in data science needs a PhD, and the only task is to come up with math algorithms and such. But, there are so many other things that you can do.

Kevin: Yep.

Mala: Describe a few things you wish you knew when you started your career that you know now and you recommend to people.

Kevin: I think the biggest thing is soft skills. I always wanted to be the guy in the back who just hacked on stuff, and that turns out to have been a mistake. Being able to talk to the other side, the people who are generally considered the profit centers, and to make some relationships, really helps you in your job. First, you're going to be able to understand what others want. You're going to better understand how you can write applications or how you can focus on what types of query tuning that is needed most.

But also, people are social animals. They work better when they have relationships. When you sit down to have lunch with people, they're no longer some abstract concept, like "the salespeople." They're humans. They have names. They have personalities. Some of them have interests. You probably won't like some of them, but yeah, this "cardboard cutout" is a person. It's a lot harder to act poorly toward a person than toward an abstract concept.

Another thing that's really important that I wish I knew early on—though I was lucky to have my first boss make this clear to me—is to make time to learn. Nobody else is focusing on building your skills. It's on you. You have to be the person who's going to be responsible for your development.

I've had great bosses who pushed this on me, but I know the norm is not, "Hey, I'm going to force you to take time and study this thing, even if it's during work time because I know that a year from now it's going to pay off for me."

Most bosses aren't going to think that way. Not necessarily because they're careless or anything. It just slips the mind. There's a lot of stuff for a manager to think about on a day-to-day basis. So, you've got to focus on this yourself.

My last advice is to get involved in the community. I did not jump in for several years. I finally started going to user groups and conferences after about four or five years as a web developer–turned–database administrator.

My first SQLSaturday was in Columbus in 2012. I really wish that I had gone earlier. I really wish I'd gotten involved in the local PASS chapter and the .NET User Group. I could always find some excuse. "Oh well, it's too far of a drive," "I have things to do that night," "Oh, I'm so tired," whatever. It's something that I really wish I'd beaten myself out of earlier. Now that I've moved to the Raleigh-Durham area, I'm much more active in the community. When I came down here, I was working from home and so, I would use the user groups as a way for me to get out of the house and actually have face-to-face conversations with people.

So, I kind of had that carrot in front of me where, "Yeah, I've been sitting at the house for three days straight. I should go out and talk to people." It took a lot of stress off my wife.

Mala: It doesn't work to talk work with family people at all.

Kevin: Nope.

Mala: What's a typical day in your life as a professional?

Kevin: So, my life nowadays is managing a team of data engineers. For the management portion, I absorb reams of paperwork. I basically knock down barriers. They have things they need to get done. We have things we need to get done. It's my job to keep outside people from bothering them.

We run as a scrum team, so we're constantly talking with our product manager, and trying to figure out what features to implement and what products to develop. There is a lot of research on this team, but there is also quite a bit of development.

On the nonmanagerial side, I'm still the database engineer for the team. We have a small team of developers and statisticians. One of our developers specializes in application development, doing what I call "statistical implementation." He's not the PhD statistician. He's the guy who, once the PhD statistician figures it out, implements the solution and makes sure that everything works as expected.

We do have a trained statistician on our team, and we also have a computer scientist with a PhD who specializes in neural nets.

Mala: Oh, nice. Wow!

Kevin: So, the three of them are different parts of that data science triangle, and I'm the database guy.

Mala: That sounds like a really cool thing.

Kevin: It is. I really like this team. They have made management easy for me.

Mala: Describe how a DBA or a BI professional can expand those skills and embrace the new world of analytics.

Kevin: The easiest way to do that, I think, is to pick up a book on stats. I mean, there are a lot of stats books. Go back to the college textbook if you have to, but there are books that are focused on statistics. There are online courses. edX has quite a few of them. You could go to Pluralsight and learn more there. And, start applying some of these things to your own life in a small way.

Let's say you're a database administrator. Something I love doing in one of my sessions is show the Glenn Berry DMV query for what your CPU utilization is by minute, over the past 256 minutes. You can scroll up and down that and get an idea, but I could throw that data into R and graph it very quickly.

Collecting this data over time, I can perform outlier detection. I can follow techniques for trending where I can see hey, we're always in this eighteen to thirty-four percent range, and yet suddenly we've jumped to seventy-five for several minutes straight. Something has changed. That might be worth looking at.

So, finding these types of techniques and just putting them into place at your job now helps give you that bridge from "I'm a data specialist" to "I'm now a statistician and a data specialist."

Mala: That makes sense. Some people say that they're not into math and that they're not math people. Do you say that there is a strong connection between doing analytics and being good at math? I've seen debates go both ways on that.

Kevin: I like to troll people and say that statistics is applied philosophy. In truth, I think statistics is more philosophy than math. In some fields, like neural networks, you have to be familiar with calculus and understand the chain rule and gradient descent, but for the most part, you're trying to understand concepts more than performing calculations by hand. We have computers for that.

That said, there is some math involved. And, yeah, for people who say, "Oh, I'm not good at math," if you want to go into the field, you'll have to get over that fear and dive in. Be able to read an equation because it's not really as bad as it seems.

You'll see a paper, and it'll be full of these formulas. I saw the same thing in economics papers. They're full of these imposing-looking formulas, but by the time you're done going through it, you say, "Oh, this actually wasn't that difficult at all. It was all really simple. You just threw in a bunch of scary-looking Greek characters."

Mala: You said a little bit about working with agile on your team. Can you get into that a little more in detail? And what are the tools you use to make it work?

Kevin: Sure. So, I guess I'll start back a few years at a prior job, because I had created the first set of tools. We had a continuous integration process back in 2010 using TFS [Team Foundation Server]. We automatically deployed applications, databases, reports, integration services, packages, services—the whole works.

The database portion was built around Visual Studio database projects. And then, I really got used to that. We were pushing things out nightly and responding quickly to user requests and needs.

We purposely chose not to implement the rules of scrum or kanban, but we had a process in place that worked well for us. My next position was at a company which did only manual deployments, and it was a big step back. I had gotten used to automatic deployments for so long, and now I had to run developers' scripts manually and make sure they actually worked.

In my current job, most teams use scrum. We have about a four-week cycle for development, and we have a continuous integration process. It's too large of an environment not to have automated tooling for deployment. Most of that tooling had been handcrafted a long time before I came in. We've got people who have been maintaining it since then—improving it in various ways and adapting as we shift technologies.

I've been a database engineer—a database developer—on a scrum team. Scrum purists hate the idea of generalists. I enjoy explaining why they're wrong—that databases require specialization. You cannot be the database guy, and the application developer guy, and the UI guy, and the QA guy, and all of these other things very well.

You can be certain parts. You could be a generalist who is okay at several of these things, but when you're in a system that actually requires saying, "Oh look, we're calling this thing 500,000 times an hour, and it's running at about one second," that's what a generalist can get you.

A trained specialist who's good at performance tuning might get you down to a few milliseconds. Somebody who's awesome at it might get you down into microseconds. Because there are so many different levels, this is like going to a hospital with a railroad tie coming out of my sides. I could see the general practitioner, but I'm going to want to see a specialist.

I consider application development in a similar vein. You've got to have specialization once you reach a certain point—once you reach a level of difficulty.

Mala: Makes sense. So, are there any tools that you currently use with the frequent deployments and such?

Kevin: We have an entire release tool which was custom-built. Within the database world, we will save all of our code in Git, but we don't use Visual Studio projects. Database projects turn out to be a little too limiting for our system.

There are so many interlocking pieces in our solution that people have tried but after about a month, they get a tiny fraction of the way, and then you say, "This isn't really maintainable because, in order to go any further, you need to create circular references," which Visual Studio database projects don't like. It ends up being more pain than it's worth.

So, the database deployments happen through this release tool and also through custom-built PowerShell scripts, which take files and run them on the servers that they need to go to.

Mala: Describe your experience with cloud deployments.

Kevin: Prior to 2013, I was very strongly against it. So, this was when I worked a state agency. I considered it to be a huge amount of risk, and if there's a security problem with the provider, it is our liability.

In other words, if I took all my data and put it up in AWS, and then it turned out that there was some sort of security vulnerability that allowed a neighbor to get my data, that's my fault. Amazon is not going to say, "Oh, whoops, that's our fault. We'll take the lawsuits." It ends up being on us.

So, prior to 2013, I really preferred on-prem. Then, I started using Azure for an ill-fated startup that a buddy of mine and I worked on. I got a taste of what that "no data center, no hardware, no problem" development looks like, and I started to understand that there were some scenarios where it made sense. I still was saying I wouldn't put a lot of very important data in there. I wouldn't put PHI—protected health information—data in, but I could see data that's just okay for the public to view.

Then, I started working at ChannelAdvisor. A lot of what we do sits in the cloud. We're often using services like Elastic MapReduce to get Hadoop clusters, or Lambda, which lets us run serverless functions. There is Kinesis, which I like to call Kafka-as-a-Service, but it is message brokering. We make considerable use of [Amazon] S3, heavy use of [Amazon] EC2, and we even tried out RDS [Amazon Relational Database Service].

Some of those EC2 virtual machines run SQL Server, too. We started using it for Disaster Recovery but I think that we as an industry have gotten to the point where there are still security risks, but this is something that companies are understanding better. We've seen enough breaches that we can price the risk value there.

Mala: Right.

Kevin: So, that said, there is still that big trade-off between OpX and CapEx, that is, operational and capital expenses.

OpX sounds wonderful for companies because there's no down payment. You get benefit immediately and you can start the bill small, at least until you realize that you made a mistake and a bill that's two orders of magnitude bigger than you expected comes in. So, that's when OpX starts making people upset.

But with that said, it seems companies are focusing more on the OpX model where they can scale as there's growth instead of buying the bigger hardware locally, making it a capital expense and depreciating it over four or five years.

Mala: That makes sense. I think that's just picking up as of now.

Kevin: That sounds right. I didn't hear a lot of it, say, four years ago but I'm hearing a lot today.

Mala: One of the much sought-after skills right now is data visualization. What has been your approach to this, and how do you recommend learning more about it?

Kevin: So, I've really gotten into data visualization over the past few months. And, I was lucky that many of the people I know really well in the community are data visualization experts, and so I got to sit down and watch them. I got to listen to them explain concepts, pick their brains a little bit, and then start learning it on my own.

There are a few really good books on the topic. I haven't picked up most of those books yet. Instead, I've gone more secondhand— reading a lot of blog posts, watching videos, getting the synopsis of the topic, and kind of digging into some of the more technical aspects.

I see visualization in two angles. There's the technical side and there's the aesthetic side. The technical side is things like the rule of thirds in photography, where if you can segment a photo into thirds vertically and horizontally, the intersections of those lines are the parts of an image where people are naturally inclined to look. Another example is color vision deficiency, where we can see what an image looks like for a person who is protanopic ["red colorblind," meaning no red cones], deuteranopic [no green cones], or tritanopic [no blue cones].

And so, you can see the technical part. This isn't how to make a beautiful design, it is more of how to make something that people can see knowing that about eight percent of my population is going to see this color scheme differently than I do. And, that's kind of the area that I've stayed in on—the technical visualization side.

I've been able to take advantage of tooling like Power BI, where they already give you a lot of this stuff for free. I don't have to design a bar chart. I can just layer on a bar chart that looks pretty solid, and then I can focus on color selection or the layout of the screen.

The next area that I've delved in to is ggplot, a library that implements the grammar of graphics in R. The grammar of graphics is an idea that I can distill from a very complex image as a series of rules, as a series of layers—one thing following after another.

Hadley Wickham has a book on the topic, and there are other great resources, including cheat sheets and tutorials, where the author walks you through fifty lines of code on designing a beautiful plot. Here's a complex plot that looks professional grade, and you can reverse back, and say, "Oh, well, all it is is a bar plot, and then on top of that, we're changing the theme, and we're changing labels, and we're resizing fonts. We're layering it on some annotations. Here's some text on the screen that helps explain this little portion of the plot and softening it up. We're removing some of the sharpness of the image." And, by the end of it, you have a really nice looking visual.

That's how I got into the technical side of visualization. I am interested in digging deeper into the aesthetic side as well. Visualization is an area that I've historically been awful at, but it's kind of interesting to slip my foot in the door of this field.

Mala: What are some of the new technologies in the data world that you are looking out for and that you think a SQL Server person should keep up with?

Kevin: First, definitely streaming technologies like Apache Spark Streaming or Apache Kafka Streams. Even if you love SQL Server—and I do love SQL Server—or if you love Oracle or DB2, there are a lot of places where you want to use a streaming technology. You can use SQL Server as the source or as the destination. But in between, let's say I have a lot of messages coming in IoT devices. In that case, I don't want to hit that SQL Server hundreds of thousands of times per second if I can avoid it.

Mala: Sure.

Kevin: I want to instead have a message broker buffering and maybe aggregating the data, but definitely feeding SQL Server data in chunks because that's generally more efficient. If I can bulk insert a million rows into a column or a table, that's going to be a lot more efficient than inserting one row at a time a million times.

So, that's an area where you easily see integration. But, you can also use Spark Streaming or Kafka Streams and take advantage of the quick retrieval and aggregation of data and feeding that out as a real-time stream of information.

When you have that real-time stream of information, things like a dashboard now get to be a lot more interesting, because I can see in real time what's going on. I don't even have to wait two or three minutes.

Mala: Oh, that's fascinating.

Kevin: Yeah, if I have a fast enough system, I can just fire off messages, put them through Kafka, and pull them back in through another application. It's like Service Broker if Service Broker actually worked, which is mean because it does kind of work but ...

Mala: It does kind of work, that's a good way to put it. What is the role that documentation and communication play in your job?

Kevin: As a manager, I probably should say lots of documentation is critical because we need paperwork and blah, blah, blah. In reality, it's reasonably important. It is most important to us when we're doing research. We want to be able to say, "Here's the stuff that we've done. Here's the conjecture that I've got. Here's how I tested that conjecture. I formed it into a hypothesis, and here's how we can validate what we're doing. Here are the things we've tried, the things that have failed, the things that have succeeded." That's important I think.

That's vital to our team because we build up a set of information, and if one of us leaves, or if my team expands and I bring somebody else in, then we have artifacts to a piece of that culture. So, there are benefits to it. But, probably a little bit less than if I were a database administrator. If I were still a DBA, I want to document everything because something weird is going to happen at three a.m., and if I have it documented, if I have a policy written out, then when I'm completely exhausted, and I get that on-call message, I can just go through the list and get the right result.

And, the more wakeful, energized me of three months ago when I wasn't stressed was able to go through and build a policy. In our team, because we are mostly a research and development type team, we don't have those three a.m. emergencies, so our documentation tends to be a little bit more geared toward research.

Communication. So, our entire shtick as a team is we build tools. We build microservices for other teams within our company to use, and part of my job as manager of that team is selling our microservices. It's talking to other teams and figuring out what they want. What kinds of things are helpful? If we build something, what kind of value would they get? What do they need? What endpoints do they need? How can we do this? What are their demands? And then we try to figure out if we can do the product.

So, that level of communication is critical for us because if other teams don't know that we've got something out there, then either they're losing out on something interesting, or they try to do the same thing. That unnecessary duplication of effort is something I'm trying to prevent.

Mala: Makes sense. What are your favorite books, blogs, and other means of learning? How do you keep up? Are there any from the non-SQL Server world?

Kevin: Yes. Confluent, the company behind Kafka, has a blog that has a lot of great information and some nice tutorials that go into great depth. Cloudera's engineering blog is also great because they work with customers, and show architecture and implementation details to help organizations get a performance benefit, or show how they've been able to scale out their solution. Instead of just giving the marketing fluff, they go into detail and show the types of tools they used, how they configured it, and how they changed things to get a much faster performance."

There's another blog I like that's run by Knoldus, a consulting agency out of India. Each of their consultants does some blogging. Some specialize in Cassandra, some in Spark, some in Kafka. Quite often, I can read through and find something interesting about Spark Streaming that I didn't know. Many of the posts start basic and build up knowledge over the course of a series of posts. That's another blog that I recommend checking out.

Mala: That's a good list. What is your style of interviewing the data professional? What do you look for in people you wish to hire. What are some examples of questions you ask?

Kevin: I want people to say, "I don't know." I'm going to push you until you say, "I don't know." I dig deeper into a topic and what happens is either the interviewee will say, "I don't know the answer to this," or they'll start making stuff up. If they start making stuff up, that's an immediate red flag, and most likely I'm not hiring you.

For the phone screen, it's about seventy percent technical questions and thirty percent nontechnical questions. Technical questions are pretty simple, like, "Can you tell me the difference between a clustered columnstore index and a clustered B+ tree index?"

And then, I go into some nontechnical questions. My favorite one is, "Tell me about a great business problem that you've solved with SQL Server."

This question makes a person think. I have had cases where interviewees stopped to think about it, and then they've given me an interesting scenario like one where the business was having a crippling issue where they had to take data out, put it in to Excel, go through all these formulas, put in an ugly macro-filled workbook, and then manually type in the results in some other application. So what they did was use Integration Services and automate pulling the data out of the first system and putting it in the second system.

That's the type of thing that I want to hear about. If I hear, "Oh, well, I rewrote this query to make it run in a second instead of a minute," that's not answering the question.

Mala: Wow! That's interesting.

Kevin: Sean McCown had a great question that I've stolen and that I use a lot. "Ask yourself a question and then answer it." That gives me a chance to see if I am missing something. It could be that this person knows a lot about merge replication, and I will admit, I ask zero questions on merge replication in my interviews. But, if this person knows a lot about it, that's wonderful. You can ask the question, you can answer it, and then I can ask you some follow-up questions or for more information, and go down a road that I otherwise would not have been able to.

It's also an indicator. If a person asks a question that they cannot subsequently answer, that's a pretty big problem. I'm even willing to accept something that's a really easy question and say, "Okay, that's not a great sign but at least you didn't ask yourself something you can't answer."

My in-person interview is a lot different. On the phone, we have plenty of give-me-an-answer type technical questions. In person, it's screenshots and conversation. I've got a slide deck full of screenshots, and all I'm asking is,

"What do you see in this? Is there anything that's interesting? Any questions you would ask about it?" and I explicitly tell them there are no right or wrong answers. This is a technique that I stole from Brent Ozar, and it works wonders.

We spend the interview time talking about what that person sees. If I show the interviewee a screenshot from SQL Server Management Studio, they might respond, "Oh, well, you've got columns, keys, indexes, and statistics," but then once they start getting into it, the good candidates start asking real questions, "Why do you have these columns? These columns look like they're pretty similarly named. This is a strange primary key."

Mala: Right.

Kevin: I can always come up with a justification—sometimes it's the real reason, sometimes it's something I made up. I want to see their thought process. I want to see where they ask questions. Are they going to say, "Oh, I think I can understand this but could you tell me the story behind it?"

And so we'll do that with screenshots from Management Studio, with code snippets, and a few other things that they'll do as part of the job. On the analytics side, I want people with the capability to research and read academic papers. People who are not afraid of those scary Greek characters.

I can train skills. I can teach you how to write SQL queries. I can teach you how to code in .NET. I can teach you R or Python. But I can't train you a desire to learn.

I can't train up a willingness to dig in to research papers. Those academic papers are where the best information is. They show up in papers first, and then six to twelve months later, that's when you start seeing them in blogs. Then you start seeing them in tools, and then in videos, and then in books.

If you've got somebody who's only going to read the books, that's about three to four years behind state-of-the-art.

Mala: That's a fascinating tip. I've never heard that from anyone yet.

Kevin: It's worked pretty well so far.

Mala: Are there any websites or any places you go to read research papers? Are they the same as the ones you mentioned before?

Kevin: So, no, different places. Basically, if you still have a university email address, there are a lot of places where you can get access to technical papers, but there are journals that are pretty good for statistics like the Journal of Applied Statistics.

Back in my economics days, I spent a lot of time on SSRN, the Social Science Research Network. But, there are stats-based journals. There are computer science-based journals like the Journal of the ACM, or IEEE journals, or the International Journal of Neural Systems. Unfortunately, for most of these, you have to pay to read the articles. But, with a university account, sometimes you are able to read the papers for free.

Sometimes you can get ungated versions where the professor will just post a draft copy of the PDF on their website. It may be a working paper but not quite the completed version. But if you're in academia, your goal is to get people to pick up on interesting ideas, and I appreciate it whenever an academic freely offers up their work in order to improve the state of the art.

Mala: Are there any conferences or technical events that you're particularly fond of and you like to go on a regular basis?

Kevin: I love SQL Saturdays. I travel around, see the world, talk to people, see them over and over. It has a carnival feel to me. So, where it's like, "Hey, I get to see you. We haven't seen each other in two weeks!" So, I enjoy that aspect of SQL Saturdays a lot.

There technical conferences that I mark on my calendar as must-go events. PASS Summit for a while has been that for me. I did not have the chance to go last year, but I have been several times.

IT/Dev Connections and NDC [Norwegian Developers Conference] are a couple of others that I really enjoy because of that mixture of development and database administration. I actually prefer going to developer conferences because developers tend to be so optimistic—"I'm going to write an app that'll change the world!" Database administrators—we tend to be much stodgier, like, "I don't want you touching production because your app that'll change the world is going to take down our system and I'm the one who has to field that three a.m. call."

So, over on the developer's side, they're so wild-eyed and full of enthusiasm. It gives me an energy kick for a few days until I realize the folly of my ways. But, yeah, I do enjoy going to those type of developer conferences a lot.

Mala: Good to know. I think you're the second person I've talked to who says they like the developer conferences.

Kevin: Yeah, and you learn some interesting things. I still learn something new.

But, even within an area as broad as SQL Server, where, yes, you could spend your entire career and miss things, there comes a point where you say, "I'm probably never going to use this." The slice of talks that are really interesting decreases and decreases over time simply due to repeat exposure.

With developer conferences, you can pick different languages. Go to an R developer conference, or to PyCon to improve Python skills, or to NDC to bone up on .NET. This provides a larger surface area for learning.

Mala: What has your experience been with the MVP program?

Kevin: It's been positive. I've been a Microsoft MVP for two years now. 2016 was the first year that I was awarded, and that was shocking. I was happy to be selected, and it has opened some doors for me.

I believe that having an MVP on staff has helped our company. Now we have two MVPs on staff. This helped us make inroads with Microsoft to a point where our company has been part of the EAP, the Early Access Program.

Mala: Oh, nice.

Kevin: And, so, we've gone out to Redmond a couple of times, worked with a SQL Client Advisory Team. For example, we got to try out production workloads on SQL on Linux before it was released.

It's been a lot of fun, and it's also helped our company a lot because we found bugs in Redmond that never made it to the product because they sprouted during this testing phase. I know other companies that have sent people out to these labs, and they get the same thing. It's a great experience. It's nice being part of that.

As an MVP directly, having access to the other people who are MVPs is very helpful because I get another chance to network with some extremely smart people and get a better idea of, for example, who the Hadoop experts in the community are. That way, if I run into a problem with Kafka, I have some smart friends able to help me think through a problem.

Mala: If you had one super power what would it be and why?

Kevin: My superpower would be phasewalking. That's the power that Kitty Pryde [Shadowcat] from the X-Men has. She has the ability to phase through things, so she can walk through walls, can drop through floors. It's not a very common power. I know on the SQL Data Partners podcast, Carlos Chacon asks this question of everybody. I'm the only person who has ever said phasewalking. And, admittedly, I would use phasewalking to troll people more often than not. Like, "Oh, I'm driving on the road. Let me drive on the wrong side of the street right in front of a car and I'll phase through them."

After the first couple of heart attacks, I'd probably stop doing that, but otherwise, it's not a very useful day-to-day power unless I want to go downstairs to the kitchen, and I don't really want to walk all those pesky stairs.

Key Takeaways

- It's a lot harder to act poorly toward a person than toward an abstract concept.

- Make time to learn. Nobody else is really thinking about building your skills. It's on you.

- Statistics is applied philosophy, or at least more philosophy than math.

- I can teach you R, but I can't train in you the desire to learn.

Recommended conferences: SQLSaturday, IT/Dev Connections, PyCon

Blogs/Newsletters to follow: Confluent, Cloudera, Knoldus

Ginger Grant

Principal Consultant, Desert Isle Group

Ginger Grant *provides consulting services in advanced analytic solutions, including machine learning, data warehousing, ETL, reporting and cube development, Power BI, Excel automation, data visualization, and training. Ginger works with data solutions across a wide range of industries, including insurance, education, health care, finance, and transportation. She is a prolific blogger at Desert Isle SQL (www.desertislesql. com), and a frequent speaker at conferences and events worldwide addressing current developments and future trends in data.*

She is proficient in creating solutions using the entire Microsoft Data Stack, which includes SQL Server, T-SQL, SSIS, SSAS, Power BI, and Azure data cloud components, including Data Factory, Data Lake, Data Lake Analytics, IoT, Event Hub, Machine Learning, and Machine Learning Workbench. As a Microsoft Certified Trainer, she offers training on technical topics. Microsoft has recognized her technical contributions by awarding her the MVP in Data Platform. She can be reached on twitter at @DesertIsleSQL.

Mala Mahadevan: Describe your journey to the data profession.

Ginger Grant: I started working as a .Net NET website developer. I did more of the back-end work, such as, hooking up the screen to the database, or whatever we were using, and calling APIs [application programming interfaces]

M. Mahadevan, *Data Professionals at Work*,
https://doi.org/10.1007/978-1-4842-3967-4_7

to pass the data. The people sitting next to me were graphic artists and UI [user interface] designers. They were really, really good at the front end. There were three of them, and they all drew phenomenally. Whenever they showed me their work from their sketchpads, I was always terribly impressed.

Dave, who had been tasked with designing the webpage, showed me one of the most amazing sites that I have ever seen. He arranged all the photos in a pinwheel, and when you clicked on a picture, it spun out to a full-sized view of the picture with a full description of it. It looked great. I said, "That's absolutely phenomenal," and asked, "Whatever gave you that idea?" He said, "Well, I was looking at the trash blowing around, and this came to mind." I thought at that point that I was going to do back end because there was just no way I could up come with such beautiful front-end stuff, as I didn't have those amazing artist skills.

Mala: Wow, interesting.

Ginger: At the time we were hooking up to an Oracle database, then I changed jobs and was more involved into taking data from a website into the data components, and then from there, writing stored procedures, doing database work, and from there it was pretty much all data.

Mala: Describe a few things you wish you knew when you started your career that you know now, and you recommend to people who are new.

Ginger: When I first started in technology, I thought I was going to be writing this kind of code, and that was it in perpetuity. What I do now is seek out something that I find really interesting, and learn about it. It's so easy to learn now all there is about technology topics. One thing that I would like to tell people is that there is no reason to believe that you will be doing what you're doing now forever. If you are a dev and you want to become a DBA [database administrator], that can definitely happen. You're going to have to put in the effort on your part to make it happen, but you're not stuck. I didn't think that when I first started out. I thought I was stuck.

Mala: Yeah. When you say "stuck," do you mean doing the same thing day after day?

Ginger: Yeah, I felt like I was always going to be doing a set of tasks at a particular kind of job. I didn't see that I would be working at any other kind of job.

If you want to get into another aspect of the field, you certainly can. There's nothing stopping you but you. Just because you don't see how you can work toward the job that you want, doesn't mean you can't get that job.

Mala: What is a typical day in your life as a professional?

Ginger: That's kind of an interesting question because as of December of last year, I've gone independent. I kind of split my day between doing work for clients and various marketing tasks.

I lump a lot of things into the marketing category, including working on my website, creating presentations, writing SOWs [statement of work], and going to events. I look at what is due when, how many hours that I have, organize the work accordingly, and figure out how I am going to make it happen. I've been working from home now for about four years, and I really enjoy it. I have kept the same routine and work every day, if I have client work or not, as there are always things I think I should be doing. I keep very regular hours, and some days they are regularly very long. But every so often, I go out to lunch and stay gone because I feel like it. I don't work that much on the weekend and try not to work very late at night.

Mala: I'm jealous.

Ginger: I generally start at seven, and I end at five.

Mala: Describe a BI professional. Expand their skills and embrace the new world of analytics.

Ginger: One thing that's great about today is there are so many people who are providing information at very little to no cost, which just means that you have to figure out the time you are going to take to learn, and how you're going to make that work for you.

Just as an example, yesterday I learned that there was a conference in Austin on data and covering a lot of data science topics. A number of people who attended have posted their slide decks on SlideShare. That's one place you can look at the content from a number of different presentations on events you did not even attend. There is also a lot of content on YouTube, which is kind of an interesting animal because much of the stuff that's out there is awful, but all of the past sessions for virtual chapters are generally out there. If you want to look, I think that those are generally quite good and very specific to the kind of work that we do. Finding the old recordings of somebody who's very well known in the industry talking about an aspect that you might be specifically interested in is a great way to learn.

One technology topic that I am spending lots of time on is new fields of data science. There are a number of courses out there that cover it, and I am working on mastering everything. The biggest problem I have is figuring out how to do everything. You need to figure out when you're going to study new stuff and how much time you're going to do it and make that happen. That's up to every individual to figure out how they're going to make that happen. I have a mentee, and I am always telling him, "Tell me what your plan is. I want to see how you're going to do this." Don't say, "I want. I want." That's just a dream. Make it a reality by developing a plan.

At the same time, it is important to come up with a plan that's realistic. Write the plan down, and then over time come up with something that you can actually achieve. There was a point in my life that I had a lot of personal issues, and I was kind of depressed. One thing that I found to be really rewarding is to set goals that I could actually hit, which I found very encouraging. Goals may be a certification, or completing a class. If you set goals that you can't possibly meet because they're unrealistic, then you're just beating yourself up more because it's something else that you didn't do. Plans really need to be positive.

Let's say that you want to be certified in a month, and life intervience and instead it takes three months. You can at least say that you achieved your goal. Maybe the time's wrong. Maybe you set it up to complete a class on Python, or initially you thought, "Yeah, I'll do it in a week," and you got busy, and you do it in three weeks. That doesn't mean that you're a failure. It means you completed your goal. Use your plans as a method of positive reinforcement to get to where you want to go.

Mala: What are some of the technologies that you're watching as interesting for the future, starting as a SQL Server person.

Ginger: Starting as SQL Server person, there have been just so many changes. The integration SQL Server with other languages, such as R and Python, in data science is a big thing. There is currently a big discussion in the industry on what is the best language. I think, over time, given a lot of things that are happening in the industry, that's going to be Python, and there are going to be people doing R in the future. I am personally working toward that being the case. I'm betting my time and my resources on that being true.

The other thing that I'm seeing in the industry is the evolution of big data. It used to be that everybody was dumping data into data lakes and keeping everything to go back and analyze. Going back, it's kind of like documentation— everybody thinks it is a good idea, but nobody gets to the point where they actually write it. What happens is you store your data, and later on, you can't figure out what you have. You've got comma-delimited records, and you don't know what they mean.

I see there is now more of a goal to plan how you are going to store huge amounts of data. I think that Cosmos DB—storing data in a modified JSON format—is going to be a large part of that in the future. It provides structure so that your data is not ever really stored in a pile that's going to be documented in a data catalog in some other place away from the data itself. I think that the data is going to be stored in a way that it can be easily analyzed, because if you don't, the amount of time it takes to perform any kind of analysis doubles or triples, or may not even be possible depending upon what you're looking at. That's kind of a big move I see in the industry—how data is stored in very big data components.

Mala: What are some of the ideas that you recommend a DBA or a BI person can present to management as far as analytics, big data, and buzzwords like that? What is practical and doable?

Ginger: What I think is doable, to start with, is initially install SQL Server 2017, even if it's just the developer edition. It's free. Install that on your computer somewhere. Ideally, you'd be able to get a dev version of SQL Server 2017 on a server at your company. That way you are building your skills toward the day when management decides to update. Start learning the best practices and how to use the new features. I'm seeing companies that are not looking at anything new, and they're painting themselves in a corner with some of the things that they are deciding to do. You do need to have some knowledge about what you're doing when you're starting in this kind of development.

One thing that DBAs need to understand with the Machine Learning Services in SQL Server is the impact of how code is going to run, where it's going to run, and how to monitor it. I've written a couple of blog posts on that topic. Another big area of concern is the ongoing issue of maintaining R or Python libraries that are used in the code.

In terms of management, if you want to do some kind of complex analysis of data, look at learning it yourself to gain some level of mastery. Then I would propose that analyzing some of your company data to find insights, and present those insights to management. Ask if they think it would be okay. It can be simple things, like the geographic distribution of all customers in relation to the amount of money that they spend with us. Something like that.

I think it is important is to get management buy off on doing advanced analysis. This also provides the ability to show that you can do this kind of analysis, and talk your way into doing more of it. DBAs need to know how to manage. What am I doing here? Do I need to have a lot of memory? Is this R code going to crash my server? It's not, by the way. What are the settings? What user permissions do I need? These are the thing that you need to know prior to doing any kinds of implementation, especially when it comes to maintenance. How are you going to maintain it?

As a DBA, how do you know whether or not people are exaggerating regarding the performance of their code? What things can you look for? Even if you don't know anything about the code, there are some simple things that you can look for. Perhaps an R code, for example, to see if the code will execute if it runs out of memory. There are some things you can do with R code to make that happen, and if you don't, it will run out of memory. As a DBA, you may not know much about R code, but you can find out.

Mala: Good thought. A lot of people struggle with whether they need to know a lot of math to get into analytics. What's your opinion on that? Is a good knowledge of math necessary, or are there things you can do to help with the process even without being much of a math person?

Ginger: Let me cut to the chase. Eventually, you are going to have to understand statistics. You are not doing math problems. I know a lot of people shy away from it because they didn't find differential equations that exciting in college or in high school. The most important thing, I think, is to look at what you're trying to do, and then look at the math that's required to understand it. A lot of times, it's understanding probability—the way that it's doing the analysis, but I don't think you need to understand a lot about math. You do need to understand a lot about statistics. I think a lot of it can be gained over time.

However, I will say that there is a real bias in the industry right now for people who have advanced science degrees, but I think that is perpetuated by the people who have these advanced degrees. Over time, I don't see that holding up.

Mala: So true. Don't you think that's going to last for a long time?

Ginger: No, because people want the analysis, and there aren't enough people with degrees to do it.

Mala: Correct. One of the much sought-after skills now is data visualization. What has been your approach to this, and how do you recommend learning more about it?

Ginger: I'm not a particularly visual person. What that means is I consider myself to be a person who needs to rely on experts when it comes to things like picking appropriate colors. There's nothing wrong with that. There's a number of different websites to help you pick out various colors.

Visualization is really a science. It's not I can draw, I can't draw. It's understanding the way that the eye and the brain work, and how those components can tell a story very quickly. I know that Meagan Longoria has done a number of really interesting series on her blog that talks about this—datasavvy.me. A lot of it comes down to how your eye works and how people learn. You don't need to be a visual person to understand that. I think it's really important to understand and implement the rules so that you can provide information in a way that people can readily understand it, but I also think that you need to figure out what you've got and what you can do with it before you get all hung up on how you're going to lay things out.

Sometimes you don't have the data needed to tell the story that you want to tell. Before you get head deep into data visualization, it's very important to have a really good understanding of what's possible to provide so you're not disappointed when you can't do it.

Mala: What's your experience with agile and business intelligence?

Ginger: I've worked on a number of projects with agile. I think that agile is important from a management perspective to determine how a project is going. It also holds people accountable in that somebody's going to ask them tomorrow, "What did you do today?", which helps the procrastinator stay on task. In that aspect, I think it's good. In terms of how it to database administration, I don't know that it's necessarily a good fit. I think that agile is really good when it comes to knowing the goal and then asking, "How are we going to get there? What's going on?" If those are the questions that you're looking to answer, I think that agile is really helpful.

If you're looking to answer how you can modify something to perform the best way possible, then that's not really something that lends itself to agile because you don't necessarily know how long it's going to take to come up with those types of answers. Maybe agile is good for reporting and tracking progress, but not necessarily for having a solution in three weeks.

Mala: I completely agree with that. What is the role that documentation plays in your job, and how important do you think that is?

Ginger: Documentation is really important to help people understand what they are paying for. It's kind of a boundary setter. If you are looking to do a project for somebody, it's very important to have a document that tells them, "This is what I'm going to do." You draw a box around it. "I'm not going to fix everything on the planet. I'm going to do this small set of things." This is a big deal in Power BI because, generally speaking, there's a lot of things you can do, but when engagements last three weeks or four weeks, you're not going to do everything. Describe exactly what you're going to do. That's what the documentation is good for.

Documentation is also very helpful in presenting your results. You always want to be able to create something that somebody else can maintain. Creating something that is meaningful to you and to you only is a giant waste of everybody's time. Clear documentation needs to include what it really does when you're done. If you are doing maintenance, it's important to let other people in on whatever it is that you're doing so that if you want to go on vacation, there's documentation that's available so that other people know what you did. Let's face it, at some point in time, you're going to want to leave wherever it is that you are. Documentation needs to be available so people don't curse you when you're gone because they have no idea what you did.

Mala: So true. Describe the importance of communication in this kind of work.

Ginger: Communication is really important. A lot of times with remote work, people don't know what you're doing. They figure that you're just sitting around Starbucks drinking coffee all day, but you need to counter that thought by communicating so that there's a good understanding of what's going on. It also provides a level of comfort to people. For example, if somebody assigns you a task and says, "Oh, that's only going to take a day." You agree initially, but upon reviewing the task, you look at it and you're like, "No way. This is a mess. This is a nightmare. There's no way I can get this finished in a day. It's ridiculous."

This is what communication is. Not only do you need to know when to raise flags and say, "What's the current state? I have found something. I can't do what I said I was going to do, and this is why." I know nobody wants to tell somebody, "I can't do it," or, "I was wrong," or that kind of thing, but the sooner you can elevate the problem and provide that kind of communication, the better.

For example, I had a situation where I didn't want to say that I couldn't do a job. I spent a good day on something that was a lot bigger than what I thought. The next day when I met with the client and I described the problem. The client said, "Well, if that's the case, then we won't even do that at all. We'll do X-Y-Z instead." It's really important to communicate what you're doing. Use communication tools like Skype for Business. Nothing drives me bananas more than people who always put their Skype set to say that they're busy.

Mala: So true.

Ginger: Telling the world you are busy all the time means you are never busy. When everything is a number-one priority, then nothing is. If you are always so busy, then you're basically saying, "I don't want to talk to you." That's all sorts of the wrong signals. Be able to block off time when you're busy. I know people who only answer email at certain times of the day, and that's fine. Just communicate what your status is so that people feel comfortable approaching you at appropriate intervals. If you want to set those yourself, fine, but whatever you do, don't just say you're always busy and only be available when someone schedules a meeting with you. That's the kind of communication that people need to avoid as well.

Mala: What are your favorite books, blogs, and other means of learning?

Ginger: I'm one of those people that look at various blogs when there's a topic that I'm interested in at the time. I like Data Science Central, SQLServerCentral. In terms of online training, I like DataCamp the best because the quality, I think, is quite good. In terms of books, I don't have any right now that I'm that crazy about. I'm kind of scattershot. I keep a list of the stuff that I look at frequently, and I put it in Excel, so wherever I'm at, I can take a look. I mentioned Meagan's blog, which I like. She's got a lot of really good data visualization stuff. Mico Yuk's got a lot of really good data visualization.

Mala: Oh, yeah. I love Mico's stuff.

Ginger: I do a lot of with Power BI. Of course, staying up-to-date on Power BI blogs is important. The Guy in a Cube videos, naturally. Jonathon Stewart has got some good content out there on data visualization as well.

Mala: What are your recommended ways of stress management and developing healthy work/life balance? I thought you said you sign off in the evening at a certain hour.

Ginger: Shutting my computer is a big deal. When the computer's shut, I'm done. I feel you need to draw boundaries as to when you will and won't do something. I'm into cooking, which I enjoy doing at the end of the day. I find it relaxing. I also find working out and yoga to be good ways to relax. When I get very stressed, I meditate. This is awful because I should do this all the time, as I find it helpful, but I don't.

If things aren't that stressful, I tend not to meditate and keep it as a last resort. I really think that working out is helpful for clearing out your brain.

Mala: Yes. I completely agree. Describe your style of interviewing a data professional. What do you look for?

Ginger: It's funny because I was recently at a Meetup where I was talking to a whole bunch of people who just graduated from coding boot camps. They were talking about their interview process, and I thought about that and what I would ask. Later on, they were talking about their computers. Based on what they said about their computers, I probably wouldn't have hired any of them. The reason why is they coded for a living and didn't know anything about the computers they were using. One girl said that she spent two months' rent on her computer, and she couldn't tell me how much memory was in it. I don't think that shows you are very involved in technology or are interested in learning things if you don't think how much memory your computer has. It is something you should know.

I had somebody tell me that you can learn technical skills, but you can't learn passion, which is what you should interview for. If you can't even bother to know how much memory you've got on your machine, I question your desire to really be engaged in the industry. If I was to interview somebody, and they couldn't come up with any blogs on the related topic off the top of their head, I probably would end the interview right there.

Other questions are nothing that you can Google. "What are the processes that you would use to develop a database? And tell me why." I'm only after the why, and I want to know what you learned in the past year and how you did it because I'm looking for people who will grow. I'm not interested in people who just want to be there, or who are just there because they've always done it.

Mala: What are your contributions to the community, and why do you recommend people be involved with the community?

Ginger: When you're working from home, it can seem remote. That's one of the great things about being involved in the community. It's important, not only for just casual conversation, but at some point in time, you're going to need help. Everybody does, and to develop a community where you can ask for it is valuable. I like to think that I am somebody who can help others. Recently, I read an article about how you really don't learn something unless you can explain it, and being involved in the community is how you can make that happen. I think that's another really valuable aspect as well.

I have heard that public speaking ranks among the things that people find the most terrifying, which I find kind of funny. A lack of good material scares me, but the public speaking part doesn't. I think that once you've practiced being a public speaker, you are a much better on-the-spot thinker.

For example, if you've recently given a talk on a topic and somebody asks you a question about it, you need to provide an answer right away. You figure out how to answer the question. After you've done this for a while, when you're meeting with your manager and he asks you to describe something, you are ready to provide a quick answer in the same way that you would do for a talk. There are things you learn about public speaking that can be implemented in your daily life all the time.

Mala: That is so valuable. It makes it easy to think on the spot.

Ginger: Perhaps somebody's going to ask you, "Hey, what do you think about this project?" If you're practiced in some public speaking methods, first you tell people what you're going to talk about, and then you talk about it, and then you repeat it. If you do that in regular conversation, it improves your communication skills no matter who you're speaking with.

Mala: Absolutely true.

Ginger: It's fun. Let's face it.

Mala: Describe your experience with cloud adoption and data warehouses.

Ginger: What is interesting about the cloud, is that there are a lot of people who are all in with cloud, and there are a lot of people who are really reticent to be involved in it. A lot of times, people are looking at the cost model in different ways. For example, I worked at a company where they decided to put up their own duplicate data center. This involved buying a building, and refurbishing the building, and adding diesel generators. Oh, by the way, if you've ever been in a building where they have diesel generators, you need to run the gas clean out of them about every three months.

Mala: Oh, my. Okay.

Ginger: This process is very loud and somewhat noisy. Developing all of the infrastructure necessary to ensure that the data can be up twenty-four hours a day is a big investment. You need two data centers, geographically spaced with duplicate power. You do have to think about that when you're looking at the cost of the cloud. A lot of people don't necessarily look at that.

The other thing that I've found when people migrate to cloud is they determine that, "Well shoot, if I've got all of this in the cloud, then I don't need people anymore." I talked to a DBA recently who was the DBA for nine hundred databases because his company figured if they're in the cloud, then they didn't need support anymore.

Mala: Wow.

Ginger: I worked for a defense contractor recently, and their whole concept of the cloud was, "No, we're not going to do it." I also have researched a little bit into some of the white hat hacking, and let's just say I was interested in the methods used for penetrating other instances on the cloud that are touted to be quite secure. The cost model has to be in a way that people are willing to adopt it.

Mala: I completely agree. Have you had cases where people adopted it without due consideration, and then they want to back off later and such?

Ginger: What I see happening is there is a new manager, and the new manager is like, "This is costing us how much per month? This is crazy." I've never been involved in rolling back, but I have heard of places that chose to do that. I've also seen where they determined that they were not going to migrate other applications to the cloud because they were oversold when it came to how much cooler, better, cheaper, and more wonderful it would be, and then later determined it wasn't the case.

Mala: That is my experience. Do you have an interesting or funny story you want to share?

Ginger: When I first starting doing database development, I did the same thing over and over. I thought I was stuck. Through involvement in the community, I realized how you can get unstuck in what you do. The other thing that I find amazing is how easy it is to learn stuff with all the information that's available online.

You want to be a DAX expert, you can do it. The information is out there—free and not. Starting out there were a lot of times I used to sit thinking, "I'm stuck. I don't know how to solve this. I don't know how to solve this." That doesn't need to happen anymore. I think my superpower is finding things on the Internet. You name it, I can find it. Having that skill has made a lot of things just invaluable. Literally, there's no end to the things that you can do if you know where to look on the Internet.

I will put a little caveat on that, and yes, I know there's a whole lot of crap out there too, but figure out the difference, and you can find and be anything, because the information is literally at your fingertips.

Key Takeaways

- You won't be doing what you're doing now forever. You're going to have to put in the effort to make change happen, but you're not stuck.

- Make dreams a reality by developing a plan and working the plan.

- People want analytics, and there aren't enough people with degrees to do it. So, the degree bias may not last very long.

- Before you get deep into data visualization, it's important to have a really good understanding of what's possible so you're not disappointed when you can't do it.

- You can learn technical skills, but you can't learn passion.

- Public speaking helps you provide a really good answer to questions that you wouldn't necessarily be prepared for.

Blogs/Newsletters to follow: DataScienceCentral.com, SQLServerCentral.com, Mico Yuk's blog (http://bibrainz.com/aof/), Jonathan Stewart's SQLLocks.net, Meagan Longoria's blog (https://datasavvy.me/author/mmarie4blog/), Power BI blogs

Vicky Harp

Principal Program Manager Lead, Microsoft Corporation

Vicky Harp *grew up around computers. The daughter of a firefighter father and a mother who ran her own computer business, by high school, she was regularly competing in coding competitions. As an undergraduate at Texas A&M University, she discovered a strong interest in database systems, going so far as to become a lab assistant for database courses to help other students with homework and normalization concepts. An internship in the data warehouse group at Compaq Computer Corp. in 2000 began her career working with SQL Server.*

From junior roles in technical support, Vicky worked up to software development, coding professionally in both web and application development roles, but always with a deep focus on the data access layers and an eye toward performance tuning. Since 2004, she has primarily worked on commercial tools for SQL Server, working her way up to the role of lead developer before transitioning to product management.

Vicky enjoys speaking and writing in the SQL Server community. She has presented at dozens of events, both in person and via webcast. She is also active with technical communities on Twitter. She considers these interactions critical to her understanding of the SQL Server tools space; industry trends, user pain points, and emerging use

© Malathi Mahadevan 2018
M. Mahadevan, *Data Professionals at Work*,
https://doi.org/10.1007/978-1-4842-3967-4_8

cases all present themselves organically in these venues. She follows industry news sources, white papers, and research from a variety of sources. While her primary technology is SQL Server, she is also excited about many other technologies, and she is carefully following the evolution of data practices in cloud-enabled environments.

Vicky now considers herself primarily a product manager and architect, but she is equally happy to jump into code or to discuss strategy with an analyst. She seeks to challenge assumptions about work/life balance in the data profession, believing that a more harmonious approach toward one's career leads to better work and to more inclusive workplaces.

Offline, Vicky enjoys fairly analog creative pursuits, such as gardening, calligraphy, sewing, and reading. She also enjoys letter writing on one of her two antique typewriters, and hiking or going to the beach with her family. She can be reached on Twitter at @vickyharp.

Mala Mahadevan: Describe your journey to the data profession.

Vicky Harp: I really kind of fell in love with data the first time that I was exposed to it in programming. I grew up around computers. My mom ran a computer business in the 1980s. She actually did some database work then. She worked with dBase and the applications that were big at the time—Word Perfect, Lotus Notes, and all of that.

I was around it and heard about it, but it's not like that's what I wanted to play with as a kid. What's funny is one of the first things that I did database-wise was in my [high school] computer science class, writing in Pascal, when I was in around eleventh grade. We were assigned to write a database in Pascal—something that would write records to disk and bring it back up from disk. Of course, it wasn't anything fancy. It wasn't relational or anything like that, but I thought, "Oh, wow! This is neat." I actually liked writing something where the interface was pretty trivial or really didn't mean anything to the usefulness of the program. It was more important to get the correct data and to be able to look it up by this field or by that field.

When I went to college, I took a database class, and I loved it. We were learning about normalization and ERDs—all the different normal forms and data integrity. I really loved that and ended up going to the professor once I had finished the course to ask if I could keep working with it. He brought me on as an undergraduate assistant, sort of like a TA [teaching assistant], and I held labs for the next few semesters. I would have office hours—labs in the computer lab—where people who were working on database homework and assignments would come. I helped them with their ERDs and their programming. I had a blast with that. I just really liked it. Even at the early educational stage, I really liked data.

That led to me getting my first internship with the Compaq computer data warehouse team in Houston. I had an internship there, and that was my first exposure to SQL Server. They were using SQL Server 7, but they were installing and looking at SQL Server 2000 at that point. I was amazed. They were showing us how they could go from the name of a customer all the way down to the exact receipt for the thing that they ordered. It was just a big data warehouse, and it's something that you and I would probably consider pretty typical now, but I had never seen anything like that before.

That was how I really fell into SQL Server. After that, as I got into programming, I tended to gravitate toward the data access layer and the database side of things—even though I was an application programmer and developer. I went to a couple different companies. I went to IBM, and they were using SQL Server. I went to an engineering company in the Houston area, and then from there, I went to Idera. Part of why I was interested in Idera is that they were doing tools. It was application development, but it was for SQL Server, and that's how I got into the tool space. I've more or less been doing that most of my career now.

Mala: How did you end up working for Microsoft?

Vicky: I was at Idera for thirteen years. I went from tech support to junior developer, staff developer, lead developer, product manager, product strategist, to product management. I had a bunch of different roles and kept going up the ladder.

I kept getting the bigger picture, more strategic, etc., but I always worked with the same customers. I decided to move on from them and try out my hand at data architecture, and I was actually very, very happy with that. I got in touch with Microsoft, and they had a need for someone on their tools team. That was a very tight fit with what I'd been doing, and so it was a pretty natural decision. It's a wonderful opportunity to contribute to the user experience for all SQL Server users, and something I'm grateful to be part of.

I really feel like, on a technical basis, I know SQL Server, but on [the basis of] industry knowledge, user knowledge, knowing the personas, and knowing the players, tooling for databases and for SQL Server is a little bit more my sweet spot.

Mala: As a product manager, you work with a lot of cross-functional teams. That's typically something a lot of techies are not very good at. What are some of the good practices you'd like to share? I know we talked about empathy, so whatever you want to say in that regard would help.

Vicky: I think that there are two different ways in which I feel that product managers are cross-functional in a way that a lot of techies aren't comfortable. One of them is cross-functional within your company, and one of them is cross-functional with respect to your customers.

I think a lot of techie folks almost make a joke of how they're introverts, or they like to stay in the basement or whatever. That's fine. I feel that same way. Socially, I'm a quieter person, and I like quieter pursuits, but in order to make software and to come up with solutions that really solve problems, you need to reach out to better understand the problem. Otherwise, you get into this situation where you're doing things because it's neat or because it's what would solve what you understand the problem to be. But you need to understand it from another person's point of view, which is where empathy comes in.

I think you have to get yourself out of the way. You can't say, "I'm technical, so if I were encountering this problem, I would want it solved this way," because then you only come up with technical solutions and solutions that are well suited to people who are just like you. If you're dealing with someone who is technical in a different way than you. We tend to act like there are technical people and there are non-technical people. It's not that simple. Let me tell you, people in marketing are very technical with things that are not necessarily software, but they know a lot about metrics. They know a lot about how to identify things that do and do not work. They understand a lot of things that the average "technical" software or data person doesn't understand.

You have to get into their mind space. You have to listen to what their needs are with understanding that they know what they're talking about. You can't go in thinking, "Because I'm technical, I'm the savior." You go in as a servant. You go in with humility, with an attitude of "please tell me about your problem."

The other thing is I talk about is "getting out of the way of the problem." You have to focus on what in product management circles is sometimes called "the job to be done," which is what they're trying to accomplish, and don't let yourself get caught up in what it is that they want to see on the screen, or what they want to click on. They may try to tell it to you that way because that's what they think you want to hear. When they start down that path, you may also want to hear it that way, but it would be better to get something that's more of a "here is the problem I'm trying to solve."

If I'm thinking about talking to someone who's using a SQL Server tool, a lot of times they want to tell me, "This is the button I want to push to see the thing." But a lot of times the real problem they have is, "I want to know what's running slowly." You can take it back a step further than that. You can say, "I want to know why my website isn't fast." Or you can take it even a step further back than that and say, "I don't want to get paged on the weekend." Those are the actual problems they're trying to solve, but what they're going to try to tell you is, "I want to have a screen that has a button that I can click to do this."

That's something that I think everyone in technology could benefit from —try to unwind the problems. Sometimes I've challenged people to ask why five levels deep. Why do you want that? And okay, having heard that, why do you need that? Almost a childlike curiosity into what it is. That helps you come up with more elegant, overarching solutions, and you can really delight people that way. Sometimes their problem can be simpler to solve than they thought.

The other thing is that when you're working with cross-functional teams, be careful in your communication. You have to balance empathy with being authoritative. People don't want to go to a doctor and have the doctor just asking you what they think would make them feel better. When people are talking to you as a technologist, they want to hear some solidity in your thinking. Even if you don't have a solution, you might at least project a sense that you're taking diligent notes. You're thinking about it in a structured way, and when you do get back to them, you're going to come to them with humility but with some sense of professionalism and certainty.

You have to balance that. You want to be approachable so they don't feel like as soon as they start talking to you, you start spewing back jargon at them. But you also don't want to be so soft that they don't understand that they're talking to an expert. I say "expert" because when you're talking a cross-functional team, even if you look at yourself against the experts in your field and feel some sense of impostor syndrome, a lot of times when you're in that conversation, you are de facto the expert in the room. You need to accept that mantle gracefully and graciously.

Mala: That's very well said. A contextual expert.

Vicky: Yes, a contextual expert.

Mala: What are the most important things you did to ensure success and value in the product you managed?

Vicky: There's an entire discipline in product management outside of being involved in data and being a data professional, but interestingly a lot of those things come down to data. You need to get information about whether people are using their product, whether they would recommend it to other people. That's one of the values that product managers love to chase down. Do you use it and you recommend someone else use it? Would you say this is something that you've taken on as your own tool?

It's important to try to get broad information and not allow yourself to be chasing the shiny object. You can have individual stakeholders who have a very strong opinion about the way something can be done, and you can ensure your own short-term comfort by doing what they say. Either that's a big customer, or it's a powerful voice in your company or a developer who really insists that this idea that they have is going to save the world. That might in the short term make you more comfortable, but it might not actually make your product better. Everything you add to the product is something that you have to maintain in the product, or you have to come up with a way to remove it later.

Really, one of the important things you can do is to say no to a lot of things and to really get comfortable with people feeling like you're someone who always says no and not taking that to heart. You have to protect the majority of your customers against the loudest of them sometimes.

Mala: How would you manage conflicts with stakeholders in the process of managing a product, at least when people want different things?

Vicky: That all again comes to internal and external. I find one of my most effective things is just almost audacious levels of honesty. Whenever I'm talking to someone and I know the data, and I have my reasons, and I feel like I can justify why something is or isn't a good idea, I try to share that information. It often diffuses a situation. If someone really thinks that something's important and you can come back to them and say, "I agree that it's important, but we have these other five things, and I feel that they're more important. Which one do you feel we should replace with the thing you're asking?"

It's pretty unusual that they would say, "I think your third thing on the list isn't that important and my first thing should go in instead." If they do, I think it's important to be humble and say maybe they're right.

There are other times that it's not nearly as smooth as that. You run into personalities, you run into people who really don't want to take no for an answer. That's something that you have to manage on a more interpersonal basis, and I think it's a person-to-person thing. I've had people that I've worked with on my team that would get very persnickety about various things. What helped a lot was to empower them to do the research themselves, so I would say, "Okay, can you go do a write-up on that for me?" Or, "Can you go find some numbers on that?" Or, "Can you give me a plan for how we're going to do that? I'd be happy to evaluate that."

I don't have to be right. They can come back with an excellent plan, and I might agree with them. Or maybe I don't have a dog in this fight, and this is between two other people. The person who goes and does the work and brings forward a proposal, they've grown in the process of doing that, and a lot of times I find that that meeting again later, it goes much more smoothly. Sometimes, especially in large groups and with people who really feel strongly about things, sometimes you just have to draw a line and say this is where we are. You can do it respectfully. People might not be happy about it, but they can at least not feel like they weren't listened to.

Mala: What's the difference between waterfall and agile methodologies? What's your preference, if you have any?

Vicky: I do not like religious adherence to either end of that. I've done extremely hardcore waterfall and extremely dogmatic agile scrum, and both of those have huge problems, especially when you're talking about the sort of commercial products that I've worked on. A lot of that extends well beyond the development organization, and that's one of the things that I don't think always gets reflected in conversations about those methodologies.

When you're dealing with an agile methodology in its purest form, when you're making fast changes, and you don't know what you're doing six weeks from now, that puts your marketing team in a really tough spot. That puts your sales team in a really tough spot. On the one hand, you say sell what's on the shelf, but on the other hand, it's not going to do anyone much good if they're trying to plan a trade show that's going to happen in three months and they don't know what to put on the handouts. You need to have a little bit more roadmap guidance than that, but at the same time, if you are strictly waterfall, and you're writing specifications and you have a really long lead time, you can be pretty slow.

I think that the joke is to do agilefall, which is maybe to say, "Here's the roadmap. Here are the things that we intend to work on." The exact order in which they're going to be worked on by the development team is maybe up to the development team, but there is an overall date by which they should be done. If information comes up in that process that makes it where those are going to take a longer time or get done faster, then you can adjust to it.

I think it's entirely different if you're dealing with a SaaS product where you can do continuous releases or a product that you can do more A/B testing. A lot of the tools that I've worked on have had more of a user-installed application end and more of an up-to-monthly release schedule. I'd like to go faster. I'd always like to get closer to the pure agile in terms of speed, but I think that realistically I also like to have a little bit more guidance in my roadmap.

Mala: I like that word—"agilefall." That makes sort of sense. Describe your experience with cloud adoption.

Vicky: More recently, I've been working on evaluating the different clouds and understanding the ways that they're used. I've done it for my own company. I've done it for the purposes of tooling and my own personal professional development. It's an interesting thing because you try to get all up to date with one of the cloud vendors, and then you turn around and read up on the other. By that time, the first one has gone and made a big change. They're really competing. They're really going hard against each other. It's interesting to watch.

I think that one of the things for me is understanding the overall workflows in which cloud is operating. It's very different if you're dealing with enterprise use of a cloud versus maybe SaaS software use of a cloud. If you're a SaaS provider and you're using a cloud as your backend, you have different needs than someone who's an enterprise user who's using the cloud for back-office type of information. You have different horizontal scaling needs. You have different regulatory needs. You have different deployment needs. If you're installing commercial software onto a managed data platform in the cloud, that's probably going to be a different process than if you're doing continuous integration and you're trying to care and feed for your own copies of the VMs with their data platform on them.

Mala: What are some of the industry trends with data that you find interesting?

Vicky: As I mentioned, I'm interested in some of the workflows. Continuous integration pipelines, I'm thinking of tools like Spinnaker, which really help you get things pushed out into containers and working with Kubernetes, etc. Really getting into some neat stuff for the code layer and the application layer for software.

I'm interested in the container story for SQL Server and for databases in general. It's an emerging technology for the data platform, but I think it's going to be as widespread as virtualization in a few years. The state of the art is being defined right now.

I really like the multimodel approaches that I'm seeing, like Azure Cosmos DB. I think that's really cool. I love the idea of being able to store your data under one model and then retrieve it under a different model and it's just your data. I love that.

Then some of the things that I've been playing with lately. I'm having fun playing with graph databases, and I'm having fun playing with Elasticsearch.

Mala: I play with graph, too. I really love it.

Vicky: I think it's great. For one thing, it seems to me that it's kind of cool that it can be almost innately multimodel in that you can say I'm going to have a document database, but some of these documents are nodes and some of these documents are edges, so you can also query it as graph. That's neat to me.

Some of the ways of interacting with graph and document data really tickle my fancy. It's really different from the relational models I first learned. I know that that is not new as far as the theory and the designs of it, but it's just something that I've been playing with, so it's been more on my mind.

Mala: How do you measure the success of a product?

Vicky: Generally speaking, I do believe in the NPS, the net promoter score, where "How likely are you to recommend this product to someone else?" is a pretty good measure of if a product is being successful. I do, frankly, also consider monetary considerations because I work in commercial products, but I don't think that that's the be all end all because some of the most widespread products I've worked on have been free. You also can't just count it by the number of people who downloaded it because what if they're not using it and you might not base it on the number of people using it on a daily basis either because that may not be the right measure for your product type. Some of the tools that I find most amazing, like Photoshop, are not daily use tools. I'm not in Photoshop very often, but when I'm in Photoshop it does exactly what I want.

Basically, whenever I'm in any way related to or owning or responsible for a product, I try to know as many of those things as possible. If I had to just pick one, I might just go with NPS.

I also really think there is something to be said for talking to people about your products a lot. That's part of why I lurk on social media so much—to try to listen in on people's problems and whether or not they organically find the solutions that I've been involved with. I don't consider it my role to get out there and try to tell everybody about every product that I work on. Maybe I will in a particular instance if I think that it'll help that person solve a problem, but that in some respects confuses my own research. I'm trying to see if people are finding it organically. I can't have whether my product is successful be based on whether or not I happen to catch every conversation everyone's having about the problem space. That doesn't scale.

Mala: Can you tell me about a time when you had to persuade someone in authority to follow your suggestion?

Vicky: This wasn't a hard persuasion. This was a fairly easy persuasion. After I'd gone to five or six different SQLSaturdays that had tempdb presentations, I came back and said we need to do something about tempdb monitoring because people are having a hard time with this. This was before I was in product management, when I was still a developer. I thought, "Everywhere I go, I'm hearing people talking about this. I'm seeing these sessions are completely full." I brought that information and it was accepted very readily, but it was accepted readily in part because I came to them with that information.

If I had just come in and said, "Hey, tempdb is hot! I think we should do something!" I don't think it would have gone anywhere.

Mala: Right, so bring data along?

Vicky: Yeah, bring data along. Exactly.

Mala: What are a few things you didn't know when you started your career that you know now and would recommend to people?

Vicky: This is going to be an eye-roller for people, I think, but documentation. It's key. Part of that is taking notes. If you are in the habit of going to a meeting and you come out of a meeting and go straight into your work and put into practice the conversation you had, that's not going to help somebody who wasn't in that meeting. It's not going to help you later when you can't remember what happened.

I went through a time where my manager would call us out and say, "Why are we having this meeting if no one is writing anything down?" You can't walk in without a plan for what you're going to talk about and then walk out having not written down anything that you're talking about. We can just go have lunch if you just want to talk about stuff with no repercussions. If you're going to try to get things done, you need to write it down.

That extends to commenting your code. You're commenting your code for your future self and for the next person who works on it. I always just assume that's me, but it could also be a new person. You can under document yourself into a corner so that you become indispensable as a code maintainer, but unable to grow into new roles. If you're the only person who knows all of the ins and outs, then next thing you know you can't do anything except being the person who knows all the ins and outs.

That's something that I didn't really understand early on. I wasn't ever somebody who was zealously guarding against sharing knowledge, but I would take it for granted that it would stay in my head and I didn't take notes.

Mala: Yeah, we all do that.

Vicky: Next thing you know, somebody's asking you a question about a conversation from three years ago. I really do not remember. It's such a relief whenever I can go and search my notes and say here it is. This is exactly what it was. Let me send you the notes.

I have one more thing. This has to do with email culture, which is that especially as you're moving up in your career, you are not doing anybody a favor by intercepting a question to somebody else. Let's say there's an email that's gone out from another department to someone on your team, and you know the answer, but it's the other person's responsibility. In a lot of cases, you're not doing anyone a favor if you happen to be the one who read it at eleven o'clock at night and you respond prior to that other person. You need to give people the chance to answer and to have their own visibility and credibility on things. You don't have to be the know-it-all.

I'm thinking particularly in a role like a lead developer. If you are very aware of what all the developers are working on in your team, that's great, but if someone's asking about a feature that one of your teammates is working on, let your teammate answer that.

Mala: What are your favorite books, blogs, and other means of learning?

Vicky: I like reading white papers and research studies and a little bit more long-form content that's coming from vendors. I do read blogs. I especially like reading the T-SQL Tuesday blog posts, and I follow along with things that are posted on Twitter and Facebook, and such. I'm not a real full-time reader of those streams, partly just from a time perspective and partly because a lot of it is useful to me mostly when I'm searching for something. I'm very grateful for blog posts if I'm looking for help on a particular topic. There are individuals who, if they have a blog post on the topic, I'm going to go read everything that they write about it. But I don't necessarily find it helpful to be reading the nuts and bolts of how to solve all these different problems on a continuous basis. It's just too much of a smorgasbord.

If I'm going to sit down and pick something to just do research and read and get myself up to date, I'm probably going to do something like go find a white paper or read something maybe even just from the broader tech news sector. I do chase down a lot of things from Twitter too.

I'm a fan of The Practical Dev blog series. It's a lot of different people, kind of like Medium. A lot of times they have particular topics that they bring to their Twitter posts, and I like to read through those.

Mala: What's the Twitter tag for that?

Vicky: @ThePracticalDev.

Mala: Are there any books—not technical, but professional broad-circulated books that you like?

Vicky: I like reading books about complex systems failures—so aircraft accidents and near misses, industrial accidents, and so forth. I've read the Columbia space shuttle report in its entirety.

I read those for a couple different reasons. One is a lot of those are more mature industries than technology, so seeing the sorts of risk management they do and the sort of post-mortems they do—it's amazing. You have something like Air Ontario 1363. I recently did a presentation on that investigation. I think it had 170,000 pages of documentation of that accident, and it was taken down to a thousand-page report. The ability of those experts to distill information into actionable points is really impressive, and it's something that I think we could learn from. The ability to turn it into a "here's what you can do to prevent this" report—one that doesn't just say, "Don't make a mistake." That is something that I think is really relevant to general stability and safety in data systems.

It's also interesting to me for something like the Columbia report because I have been working with tools and monitoring tools for such a long time. Those are examples in a different industry, obviously, but it's an example of the kind of monitoring that you need. In those reports you can see a list of all the data points that they got from the ship as it was failing. It shows all the ways that they did and did not know things were going wrong. You can see that in Three Mile Island too. A lot of that story comes down to a sensor that was wrong because it indicated whether a release valve had been told to open or close. It didn't actually say whether it was open or closed. That sort of lesson learned is super-relevant to informational dashboard design for monitoring and SQL Server and tools in general.

Mala: What are your recommended ways of stress management and developing healthy work/life balance?

Vicky: I read something, and I can't remember the source of it. I apologize to the author. Change the term from "work/life balance" to "work/life harmony," because if you really think about it, life should always win in the work/life balance. The idea that you're going to have a fifty/fifty arrangement between those two is not correct. Sometimes work is going to be taking a lot more of your time or attention, and sometimes it's going to be taking a very tiny bit. I think harmony is a good way of phrasing it. You need to make sure that you're not going beyond your own one hundred percent at the end of the day. That means that if your life is going in a crazy way, you need to be able to manage your work so that it's not pushing you to the point of burnout and exhaustion.

For that reason, I do believe in keeping human hours. I'm not a big fan of gold plating and putting in a lot of extra hours. I think if you can't get the work done in regular business hours, and that continues on for a while, it's an indicator that either you're not as effective at your job as you could be, or you need more help. Either of those are things that you can act upon. People are embarrassed or ashamed if they can't get things done, but I try not to be. I try to be forthright and say, "I have a list of five things to do. I'm only going to get three of these done at the current rate, and we need to decide if I'm going to change my expectation to only get three of these done, or if we need to get me working faster. What do we need to do?"

That's a way that I think about things. As far as stress management, I try to get out a lot. I go for walks. I believe that my job is thinking, not typing, so if I have a hard problem, I may best solve it on a park bench. That doesn't mean that I'm just going to go off and watch a movie. I'm going to try to do something that allows me contemplative time. At home, I switched to a standing desk. In part because it makes it easier for me to pace, and to walk away, and to just kind of walk back to my desk whenever I do have something to type.

For years and years at Idera, we were in houses. We weren't in an office tower at first. There weren't really enough meeting rooms for everyone, so we got into the habit of going to walking meetings. You'd walk around the block with whoever it was you needed to have a talk with. It was so great, and I'm still a big fan of that. Get some sunshine on your face. It makes you feel better.

Mala: Did you see that thing about the treehouse in Microsoft?

Vicky: I have seen that, and I've been to the tree houses at Microsoft. They're beautiful. In Houston, I used to go to a place called The Menil Collection, which is an art museum. I'd go there to look at the art, or I'd go out to the park outside of it and lie on a park bench and think about things, and then go back and try to solve the problem that way.

If you're sitting at your computer and your mind starts wandering, sometimes it wanders in unhelpful directions, like to shopping or bill paying, and you wind up trying to get other chores done. Maybe you're really successful at getting that other chores done, but you didn't actually make any progress on the problem you were trying to solve. And I just think it's good to give yourself the space to do that and say to yourself, "I've got to think about this thing."

I've had really tight deadlines before, where I took everything I needed and I printed it, and I went over to my writing desk to hand write my comments rather than typing them in the Word doc. It's funny, but in a way, I try to get away from the computer whenever I can. I'm also a fan of yoga for that reason.

Mala: What are the types of people you enjoy working with and why?

Vicky: I like having a wide variety of people that I'm working with. Variety is the spice of life. Young people, old people, people of various abilities, both technical and physical, intellectual. People of different backgrounds. I really like having the variety of that, but for the most part, I want to be working with people who have a helpful mindset and a desire to be helpful. I won't even say that they have to have a positive mindset. Some people just kind of have a…

Mala: More pragmatic mindset.

Vicky: Pragmatic mindset, exactly. They don't need to be cheerful about it, but just a desire to be helpful and to not be unhelpful is good. I wouldn't even say I need people to be low drama. Some people get very excited about things. That's fine, as long as they're not getting excited for the sake of getting excited.

Mala: What are your contributions to the community and why do you recommend people doing more with the community?

Vicky: I was very much a lurker in the community for a long time, and I had a manager who brought something to my attention that I completely agree with. She said that, think about the community like a soup kitchen. You can contribute to a soup kitchen by donating your money to the soup kitchen, and that is undeniably helpful. But if you go and you serve the soup and you make eye contact with people, you are going to get more out of it yourself, and the other person is going to have more of a feeling towards you, and it's going to be better … It's just better. It's not like there's anything wrong with just providing money, but going and doing the service is just going to lead to more joy.

There's nothing wrong with being a member of the community who just consumes and maybe provides by either sponsorship or by just being a reader or just by creating tools or scripts or what not, but the more that you give out to it, the more that you're getting out of it. You build friendships, and you build your own professionalism, and you learn skills that you didn't know that you needed to have.

I think that it's been really, really good for me to step outside of my comfort zone. I recommend that people do that. I absolutely understand that it can be intimidating to try to put yourself out there and feel like maybe you're asking silly questions or even to feel like you don't have any questions to ask. To feel like you understand what's going on, and don't have anything to ask, but maybe you have something to add. That's one of the ways that I think Twitter is an easy entry point, but going to SQLSaturday and volunteering is great too. The first SQLSaturday I ever went to, I just worked at the registration table. I didn't have anything to say technically yet, but I did have something to give in terms of my time, and then I ultimately started contributing a little bit more.

Mala: Are there any conferences or training programs that you particularly enjoy?

Vicky: SQLSaturday, for sure. I've enjoyed PASS Summit, absolutely. I've enjoyed the IT/Dev Connections conference. I've done that a couple of times. I've benefited from training from several of the professional training providers—SQL Skills, Kalen Delaney, and Brent Ozar and his team. I enjoy Pluralsight and Coursera.

Actually, not so much on the data platform side but in terms of general professionalism, I've enjoyed some of The Great Courses. I did a negotiation course through that company. It's available through my library.

I actually take a lot of continuing education classes in a variety of subjects, so I've taken anthropology. I've taken contemporary Islamic studies, Bollywood dancing, and French. I just try to change things up.

One year I had a New Year's resolution to spend three hours a month doing something I had never done before in my life, so every month I had to spend at least three hours on something I had literally never done in my life, and that was really great.

Mala: Is there an interesting story or something you'd like to share to wind up?

Vicky: I have severe scoliosis, so when I was a freshman in high school, I needed to have back surgery to correct it. This was all done when I was about fifteen.

It was very painful. I could not get comfortable, so I spent a lot of time on my computer. That's when I really got more into writing code and not just using applications. I'd say that the three months or so that I was recovering from that back surgery really set me up for going into computer science and going into computing generally.

I also want to mention something about empathy. I was talking with a colleague of mine the other day, who is also a woman in tech. We were talking about how working in tech and working as a woman in male-dominated fields. You can feel the need to adjust your norms a little bit. I am okay with adjusting my personal sensitivity and making myself a little tougher—maybe a little more willing to let things bounce off of me, but I'm not willing to adjust my level of empathy. I'm not willing to become someone who doesn't care or that is hard or hard-edged.

I think that that's something that a lot of people—I wouldn't say just women, but a lot of people feel that they're supposed to harden up, especially as they become more senior. I don't think that's the case. I think it's okay whenever you have a hard thing happening at work for you to really feel it. There's nothing wrong with having those feelings.

Key Takeaways

- You can't say, "I'm technical, so if I were encountering this problem, I would want it solved this way," because then you only come up with technical solutions and solutions that are well suited to technical people.

- You have to protect the majority of your customers against the loudest of them.

- If you're the only person who knows all of the ins and outs, the next thing you know, you can't do anything except be the person who knows all the ins and outs.

- Change the term from work/life *balance* to work/life *harmony*, because if you really think about it, life should always win in the work/life balance.

- You can contribute to a soup kitchen by donating your money to a soup kitchen, and that is undeniably helpful. But if you serve the soup and you make eye contact with people, you are going to get more out of it yourself, and the other person is going to have more of a feeling toward you, and it's going to lead to more joy.

Recommended training/conferences: PASS Summit, SQLSaturday, Pluralsight

Blogs to follow: SQLskills.com, Brent Ozar's blog, T-SQL Tuesday Blog Series (#tsql2sday on twitter) @thepracticaldev, The Practical Dev blog series

Kendra Little

Product Evangelist, RedGate Software

Kendra Little *is currently Product Evangelist at RedGate. She is also the founder of SQL Workbooks (sqlworkbooks.com), where she teaches performance tuning in interactive webcasts, and offers monthly SQLChallenges and recorded video sessions. She is a Microsoft Certified Master in SQL Server, and she has worked as a database administrator for companies ranging from small startups to Microsoft Corporation. As a consultant, Kendra has helped clients around the world to tune their workloads, solve their toughest performance problems, and scale their SQL Servers. Kendra is well known at SQL Server conferences for making complex concepts relatable and understandable through art and engaging demo scripts. Kendra has a master's degree in philosophy from Fordham University. She lives in Portland, Oregon, and is unashamedly obsessed with dogs and cycling. She can be reached at on twitter at @kendra_little.*

Mala Mahadevan: Describe your journey to the data profession.

Kendra Little: Like a lot of people who work with databases, I found how much I liked it by accident. I was very lucky. I was working on a graduate degree studying philosophy. I was fortunate that I had a work study job where I worked on the databases for the graduate program. They were Access databases—not the loftiest, most technical databases in the world, but I really

M. Mahadevan, *Data Professionals at Work*,
https://doi.org/10.1007/978-1-4842-3967-4_9

loved that job. I looked forward to going to work. I found all the projects really interesting. I always wanted to do more, and I'd have ideas. "Would it make our workflow easier if we could combine the data in new ways? What are the different ways we can automate the processes so there's less manual work for the people working in the graduate office?"

After I completed my master's degree, I was able to pick up a little more work as I looked for a full-time job. It was just wonderful, because it was really what I wanted to do, and I didn't have the easiest time finding a job in technology with a master's degree in philosophy. I kept edging my way closer to SQL Server throughout the next few jobs I got. That was the way I got into databases—by getting a master's degree in philosophy.

Mala: How did you end up with Microsoft?

Kendra: I ended up with Microsoft years later. I really wanted to be a SQL Server DBA. At that point, I had had jobs where I could write T-SQL, queries, and reports, but I was still not fully in charge of SQL Server databases, and I really, really wanted that. There was a contracting job at Microsoft. I think the job title was something like support analyst. The job title was not database administrator, but I didn't care because the job involved taking care of large SQL Server databases that had a lot of data to process every day, and it was so exciting to me.

During my phone interview, I asked a question: "What's your pre-production environment like?" I didn't want to work at a job without a good pre-production environment because I was used to having a good change management process. I got the contracting job, and I later became a full-time employee on the team. That was my first real DBA job.

My manager said later, "Nobody else ever asked me that question. It made me think, this is a person who cares about deploying changes properly and wants to do it right." I was happy to be that person, and that was the first time I realized that the questions that you ask as a candidate say a lot about you. I hadn't realized that before he shared that with me.

Mala: You were among the people who got to own your business really soon. Describe your experiences with that. What would you advise to women who wish to do the same thing?

Kendra: It is tricky running your own business. One of the things it's taken me some time to realize about myself is that I tend to minimize and underappreciate my own achievements. That wasn't always that big a deal to me. I was always focusing on doing better. In school, if I got a good grade, I'd think maybe I could even get an even better grade or a good grade in a harder class. I was always looking at the next thing, and not celebrating, or even recognizing, what I did well.

When you're starting your own business, not recognizing what you do well is a huge problem. When you succeed at something, it is a great thing to make a case study of it, or to share with your customers or potential customers in a positive way whenever possible. When you're running your own business, you need to really celebrate your successes and be open about them and about what you can do for your customers. It's good for your business to do that!

If you have this tendency to minimize your own achievements, you need to recognize it and try to reverse it. You need to train yourself to recognize even small victories, and act on them. Write a blog post about it. See if you can get a quote from your customer to use online or in marketing materials.

Similarly, when you run a business, there are a lot of mistakes that you're going to make. They'll be small, big, or just things that don't work out. You'll be constantly in the process of saying, "How can I change things to make this better?"

It is extremely valuable for your mental health and your company's health to practice not taking failures personally and feeling them deeply. If you're like me, you'll have some difficult feelings when you hit problems, but you can train yourself to recognize these feelings and challenge them. "I feel badly that this didn't go as planned, but I don't have to get stuck in those bad feelings. I'm going to look at how I can best move forward and focus on that." You have to really be forward-looking instead of internalizing past errors, because as a business owner, you need to always evaluate. "What is the future of my business, and how can I make the most of my work time?"

I advise anyone starting a business to think about their mindset in these areas.

Mala: You are a really creative person with what you do. What suggestions do you have for people who do a lot of repetitive work? What can you do to get more creative?

Kendra: This question is fun. I've struggled with this a lot. I am always trying to find more ways to do it. My biggest reminder to myself is to stop working and get away from the keyboard more, because most of my best ideas about my business happen when I am at spin class or walking my dog. They don't happen when I'm sitting at the computer. For me, there is something about attending a spin class and being on the bike that gives me a clear mind. I'm able to stop consciously thinking about things, and that seems to free up space in my brain for me to say, "Oh, here is something I can do." Then, after class ends, I have to hurry to write it down before I forget.

The more hobbies you have, and the more time that you can spend doing something immersive is good for creativity.

In the past, when I was working for other people, I felt that I needed to get something done because I said I would do it. Looking back, I could have asked for help from other people. I could have changed prioritization. I could have changed schedules. I could have just explained, "Here is the point that we are at in this process. This is taking longer than expected. This is when we're going to have it done." There are many times where I didn't have to try to be a hero and get the stuff done. I could have said, "I'm going to have a hobby that I do in the evenings. I'm going to get out of this office and do things," and that would have made me more creative. It would have helped me avoid burnout.

Mala: How important are teaching and presenting skills to a DBA? What would be your advice to the DBAs in this regard?

Kendra: I recommend that DBAs learn to present for a couple of practical reasons. Before I was a DBA, when I worked in technology at a software company, we often had what's called "post-mortems." We didn't like the term because it sounds so terrible. When something goes wrong, often you have to get together with management and your peers to explain what went wrong. You have to explain, "All right, this is what went down, why it went down, and how we're going to keep it from happening again." Those situations can be sweaty and tough, even if you only having to do this with your boss. A lot of times in technology, you have to explain to a customer or to a large team. The more skills you have in presenting and teaching, the less stressful these situations are.

A lot of times, there's friction between different teams at work. Having presenting skills helps you calmly explain, "Here's what we're going through, and we need to know how we can help you without it hurting us." It makes those conversations much easier as well.

As an early DBA, we would have regular meetings where different people on the team would take five minutes to discuss a project that they were working on or something that they were trying to learn, and say what they learned so far, what was difficult, the questions they were looking at, and where they were going to go with it. Sometimes it was about work projects. Sometimes it was about a new technology the person was trying out. But it was really interesting. It gave you that experience of explaining things in front of people, asking intelligent questions, and then listening to questions. That's really what you need. Doing that regularly helps.

I also, I had a group of people who needed to learn T-SQL to do their job a little bit better. It would help them be more effective if they could query some databases. There were production databases, so they needed to have some skills in order to do that. I said, "Hey, maybe I could show you some example queries and you could ask me questions about them, and we could write some queries as a group together." That was probably the first class I ever taught, and it was great. I loved it.

Little things like that can really help build technical skills. Anything that you are good at that another group might need, you can offer. It helps for when you talk to external customers of your company. Sometimes even DBAs get thrown into that mix.

Mala: What's your experience with agile methodologies, if you've had any?

Kendra: I worked with Agile way back when it first became a buzzword. I worked at this software development company where everything had been waterfall, but they switched everything to agile very rapidly, and we had very large databases. This was maybe 2004. They started deploying changes much more quickly. With these very large databases, we had to figure out how to adapt to it. Every change had to have a rollback plan, and that's actually a very difficult request for a lot of changes because if we're deploying a change to live data, by the time we go to roll it back, things may have changed from the point when we deployed it.

We began to learn about both a rollback plan and then also what I would call a "roll sideways plan." If the change doesn't go well, is there anything that would prevent us from undoing it? If so, what will we do at that point is important to know.

We would sometimes say, "Okay, things went wrong. We can't roll back, so we're going to roll forward." We had to learn different processes to do that. Having developers and DBAs on call became necessary to make things work.

Mentally, that required a big shift in the DBA team to be able to tolerate that level of change without it being super stressful. There were quite a lot of times when things didn't go as planned, especially at the beginning. We did develop more and more ways of testing things in pre-production, but we had to become more flexible when things went wrong, partnering with other people and seeing the issue through, and then just having more backup on the DBA team. Or in some cases, we learned to hand off to another DBA even while troubleshooting was going on so that someone could go pick up their kids, for example. We had to learn to work together much better.

I started attending development team stand-ups. It got easier over time because we had some developers who were more interested in databases than others, and who knew more about it than others. Some of them would volunteer to help other teams.

Our development teams were actually willing to share people with different expertise. It wasn't just the DBAs who had interests in how to safely change and manage large data changes in these environments. Finding ways to break down that barrier between the DBA team and the developers, and having DBAs interested in development, and having developers interested in DBA stuff started happening over time, and that really made it work.

In a complex environment where you're deploying lots of changes, you have to figure out things like: can we do this without an outage? If we need an outage, can we stage the data ahead of time in a non-invasive way, to minimize the time needed by the outage? Figuring that out really does require a meeting of the minds across traditional job boundaries. These days, some folks just do it with DevOps, but a lot of times, we have a specialist in database administration around as well.

I think they will continue to have them. It's complicated enough to manage your database. But I think there's going to be room for specialized administrators for quite a while.

Mala: What sort of struggles have you faced with managers and management on the business side of things, and what's your recommended approach to handling that?

Kendra: My biggest struggles have been with navigating tough conversations without allowing those conversations to be exhausting or personally taxing.

As a consultant, sometimes folks brought me in specifically to have a tough conversation. They had a conflict between different areas of the organization, and they wanted me to do an investigation and make a recommendation. Sometimes, it was clear that whatever recommendation I made, there was going to be someone who would really disagree with it, and they might have a very high-level job. As a database administrator, there could be tough conversations with other teams, or something is going on that just seems really risky. There are times where I might need to say, "I think folks aren't aware of the risk we're taking if things go wrong with this change, and we need to change the plan—and potentially change the dates we deliver this." That can be a tough conversation when release dates have been published to customers.

Anxiety around conflict makes this difficult. I've been able to handle the conversations successfully, and even have some fun while doing it—keeping it as lighthearted as possible, including different perspectives, making strong recommendations when needed. But afterward I've often felt really, really tired. I tended to be very self-critical. Being that tired afterward can make you dread these conversations. I started to realize that this reaction to the conflict wasn't due to the other people in the conflict, and it wasn't due to the situation itself. I started to realize that I could learn techniques to make myself calmer and make these situations easier.

I now have practices I can do that make me feel calmer and more centered, so that going into a situation where I need to have a difficult conversation or I need to take criticism, I have less of a response. Things like practicing meditation, regularly exercising, and walking my dog even when the weather is bad. Also, I have a process for handling negative thoughts that helps me challenge the thoughts instead of believing or accepting a load of internal criticism.

Mala: Interesting, and I really like the tools you suggest to handle that.

Kendra: I also have a network of friends outside of technology. This is nothing against people in technology, and it's hard as an adult to build up this network of friends. I have found them by going to a gym where I can meet people. There is something very calming about having conversations that have nothing to do with your work.

It just gives you this view that the world is a much bigger place, and that the problems in your work are just one part of that world. They're less challenging.

Mala: What's the role that documentation played in the DBA jobs you had?

Kendra: I think documentation and communication were how I built my career as a DBA, especially in the early days. I've always been one who learns by writing things down. I've been drawn toward documentation as a way to help learn and assimilate information. It was always a tendency of mine to want to build documentation because I'm better at remembering things if I made a document about it. This also allowed me to offload some tasks in my jobs. When a new opportunity in my group became available, I could free myself up quickly to take it on if I'd already documented most of my regular tasks I can give them an intro to the documentation, and then they can take it over. I can go try something new.

Documenting and automating yourself out of job are really useful. I have always found it to be true that they don't get rid of you. If you can fully document and automate your job, your company will find something else for you to do. I have never run into the situation where a manager said, "We don't need you anymore because you simplified the work we gave you." I suppose it's possible, but I have not found it yet. Folks usually get excited about the prospect of what else you can do.

Mala: What are your favorite books, blogs, and other ways of learning? How do you manage to keep up?

Kendra: My favorite method of learning is by writing or running demo code. If there is something I want to learn about, I love to find a way to be able to do it at least on a small scale. Right now, for example, I'm looking at some of the graph processing technology that's come out, and I really want to start playing around with it and getting my hands on it.

Sometimes finding a book is difficult because the book was written about the technology before it came out and is out of date already, or the technology is so new that there aren't any books. More and more book authors are turning to other methods of publishing because it can be so slow to publish a technical book on a technology that's evolving.

A lot of times I look for online demo materials and blogs. I won't always know what the bloggers are for a new topic. The first person who I thought of for learning graph data structures was Louis Davidson, who wrote great books about physical database design in SQL Server. I've seen some of his presentations about modeling hierarchies, and I know that some of it involves hierarchy. It turns out he has a bunch of free resources you can download, including a whole presentation on hierarchies with sample code.

Mala: Describe your style of interviewing a DBA. What do you look for, and what are some examples of questions you like to ask?

Kendra: I don't know if people like my style of interviewing, but it is described as "relentlessly cheerful."

A lot of my interviewing is asking people to think through how they'd solve the type of problems they'd encounter in the job. I try to explore what their experience has been, and also how they approach tricky situations. I ask a lot of follow-up or "probing" questions—I will not just stop at "the answer is this fact," I want to know, "Why did you think that? Can you think of situations where that wouldn't be true or if is there an exception?" I really like to dig in and challenge the candidate, but I try to remain as positive and as supportive as I can be during the process. I don't think that's the only way to interview people, but that tends to be my style.

Mala: What are some of the nontechnical traits that you look for in the person that you are interviewing?

Kendra: There are some of us who feel that we have to know the answer. Saying, "I don't know," is very hard for those folks. That is a really difficult thing when you work in technology. It tends to make the problem worse if you can't ask for help. If you have the tendency to fake knowing something, or you aren't able to question your own assumptions, that can get you into trouble, and your company can end up really suffering for it. That's probably the biggest nontechnical thing—an ability to say, "I'm not sure." Even in an interview that can actually be a very positive thing to say. "I don't know, but here is how I believe I would figure it out," is a really great thought process to share in an interview.

Mala: Absolutely. What are your contributions to the community, and why do you recommend people be involved with the community?

Kendra: I write a free quiz every week, and I write blog posts regularly. These days I also really like to draw free posters and desktop wallpapers. I like to illustrate things.

Mala: I saw that. Those are so cute.

Kendra: Honestly, thinking about this question, at first, I was like, "Oh, I don't contribute that much to the community." I thought that simply because all of these contributions are also good for me.

Blog posts are good for the community. I love other people's blog posts, but I also love my own blog posts because a lot of the stuff that I blog acts as a reference for me later. It can be good for your personal brand long term. For anyone working in technology, blogging can help you land a job. Employers will not be able to make a lot of assumptions based on your résumé, but if you have a blog, even written once a month for two years, you have twenty-four posts for them to look at, to see how you solve problems, how you write about problems, and how you communicate. That can be absolutely huge.

When I do webcasts for user groups—yes, that's a community contribution, but from doing that webcast, I get feedback from the people attending. I can learn from their questions. Their questions can make the content better, and I can also figure out what isn't clear about my presentation, what parts of it are the most interesting to them, what parts of it aren't the most interesting to them. I can also see how attentive people were during the free webcast at different points—very, very useful information for me to have, but also a good contribution to the community.

Even desktop wallpapers—which do people seem to like the most? That can be very useful information. Interacting with people and interacting with a larger community can also bring more enjoyment to your life than only interacting with folks at the office, as long as, of course, what's private at work remains private. We don't want to break that confidentiality, but having a bigger technical world to interact with makes it more enjoyable.

Mala: Are there specific people on Twitter or blogs that you follow?

Kendra: Sure. I was actually thinking about this the other day. There are a couple of vendors who are involved in this actually. SQL Server Central over the course of my career has published really useful things. They've done things like their Stairways series. They have people who know a lot about something—this free "stairway" to learning in-depth for SQL Server. It's just amazing. They also syndicate a lot of people's blogs. Sometimes I just like to dip on their RSS feed. Redgate's Simple Talk is also really useful. That's in-depth articles. SQLPerformance—I like their site for curated performance content. If I want to blow my mind on something and go really deep on it, I find an article on there.

Mala: Do you have an interesting story to end the interview with? Something funny or something you'd like to share.

Kendra: At one point when I was in college, I had to be about twenty, I suddenly had a very strange childhood memory of some kids driving by in a car and throwing a cake at me when I was around nine years old. I asked my mom if it was a dream.

My mom said, "I don't remember anyone ever throwing a cake at you, Kendra." I thought, "Wow, it seems so real, but she's my mom, and she would know."

A year later, I was visiting my older brother, who turned to me and out of nowhere said, "Hey, do you remember that time when you got caked as a kid?" I was shocked. It really happened! My brother continued, "Dad was watching us, mom was out doing errands, and some teenagers threw a grocery store cake at you while you were sitting on your bike. You freaked out and Dad had to hose you off in the side yard."

It turns out, Dad saved the day before my mom came home. Dad got all of that cake out of my hair and clothes, he dried my tears, he calmed me down, and he washed my bike. Now when I look back at this memory, it's a crystal-clear moment of my childhood. I can see lots of details of the house we lived in, I remember how much I loved that bicycle, and I can feel how connected I was to my parents and siblings.

The reason that this story matters to me is sometimes now when I go through a difficult time, I try to take a moment and realize, "Okay, even if this is really upsetting now, possibly someday I will be able to see the humor in this, or this will be meaningful for me to remember. If I work through this, I may learn something." That story has a happy place in my heart now.

It's the weirdest thing. I don't know if I would remember the same amount of detail about that time in my childhood if it hadn't happened. I don't even know if I would remember the bike. That's the story in my life where I try to remember that the tough stuff is worth getting through and can be meaningful to us later on.

Key Takeaways

- In a job interview, the questions that you ask as a candidate say a lot about you. Always prepare questions to ask in advance

- If you are running your own business, you need to really celebrate your successes.

- Focus on learning from failures, and surpassing the negative feelings around them.

- You get to choose your own adventure. You get to choose how you describe your experience as a generalist or a specialist, and that helps you go on into the future.

- Put boundaries on work—that is the first step toward getting to places where you can have creative thoughts.

- Practice public speaking. It will make you more relaxed for tricky conversations, like explaining why you should get a raise, or recapping what went wrong in a major incident.

- Accept that you're having a negative thought, but don't necessarily always believe the thought.

- Documenting and automating yourself out of job usually leads to a promotion, or more interesting work.

- "I don't know, but here is how I believe I would figure it out," is a great thought process to walk through in an interview.

Favorite conferences: PASS Summit, Microsoft MVP Global Summit, SQL Bits, developer conferences

Blogs to follow SQLServerCentral.com (the Stairways series in particular), Simple Talk (www.red-gate.com/simple-talk), SQLPerformance.com

Jason Brimhall

Principal Consultant, Database Masters

Jason Brimhall *is a Microsoft Certified Master/ Microsoft Certified Solutions Master and Data Platform MVP. He has more than 20 years of experience in the technology industry, including more than 10 years with SQL Server. He earned a bachelor's degree in business information systems from Utah State University. One of the highlights of his career was co-authoring two books in the SQL Server T-SQL Recipes series. He is a frequent presenter at SQL Server events worldwide, including SQLSaturday and user groups.*

Jason enjoys finding that magic turbo button to make queries go faster, which ultimately leads to huge relief and satisfaction for the client. He has done this sort of work for clients of all sizes throughout the world. He is well experienced with multi-TB databases, as well as environments with thousands of SQL Server instances to manage. Jason writes his own blog, SQL RNNR (jasonbrimhall.info), and he can be found on Twitter via @sqlrnnr.

Mala Mahadevan: Describe your journey into the data profession.

Jason Brimhall: It's actually rather simple. My journey started when I was working help desk, and I knew that I didn't want to work help desk my entire career. Help desk was not really my end all, be all career goal. Working help desk is a nice starting point. So, at that early age, I knew that I wanted to do something different, and do something a little more complex. I saw this thing called SQL Server 6.5. It was super hard, and seemingly ninety percent of the people taking the certification exam failed because it was hard. I decided to do it.

So, that's how it started for me. I saw this thing that was basically unattainable and super difficult for most people, and that's what I wanted to do.

Mala: So, you took the exam for 6.5? And got through it?

Jason: Yeah, I passed my certification exams for 6.5 and then started finding one-man-shop type of network admin positions that allowed me to do both the SQL administration and the network administration. At that time, businesses were trying to go in the direction of consolidated smaller teams, so they were hiring for jack-of-all-trade administrators instead of specialists. I didn't want to do jack-of-all-trades for very long. So, I worked hard to try and convert that type of role into a specialization a few years later.

Mala: Describe a few things you wish you knew when you started your career that you know now and you recommend to somebody who wants to be a DBA.

Jason: That's actually kind of a hard one because that's all part of the journey. If you knew it then, you'd take it for granted, and you may not necessarily get to where you are now. While it might be nice to sit down with yourself from years ago, I don't know that I would necessarily offer myself advice about what to change or do differently, except maybe to be more aggressive in the public arena, such as in the community/family type of setting that SQL Server has. That might be the one thing, but again, I wouldn't say that you have to do it. I would probably try to just leave little hints (to myself) that you want to do this sooner rather than later. In my opinion, your growth, learning, and understanding really starts to take off when you start to share more. So, I might just try to convince myself without being overly obvious about stating what I'm trying to get myself to do. Does that make sense?

Mala: It makes total sense to me, because that's one thing I didn't do early on, and I wish I had. Do you have any experience with agile with regards to database administration?

Jason: Practically everybody is doing pretty much Agile or DevOps, and it's all the same shtick for me with just a different name.

Mala: You don't see a difference between Agile and waterfall approaches with regards to database administration?

Jason: If you're doing your job correctly, no. What one place does may be called something entirely different in another place. For instance, you can call it "waterfall" in one place, but what you're doing there is "Agile" in another place. The only real difference between them is how frequently you decide to release something. To me, it's all the same deal. People just decided to go with "buzzword bingo" names and try to hop on a bandwagon because it's the hot topic of the moment. But the only real thing that matters from all of them, in my opinion, is how well the team communicates. If they can't communicate with each other effectively, then your methodology, your ideology—it's crap at best.

Mala: Describe your experience with cloud adoption.

Jason: I've had quite a bit of experience with cloud options from implementing new options, to migrating everything entirely to the cloud. Whether it's related to big data or it's transactional, or whether it's just trying to get the basic facts for people, or even clients investigating MySQL in the cloud, things like that, to me, the cloud is nothing new. It's nothing surprising. It's nothing different. It's just a buzzword bingo name. The fact of the matter is that it's still on hardware. It's still in a data center.

The only question is what data center is it in? Before cloud went big, most of my clients had their servers and their data in a shared data center somewhere anyway. They didn't call it "the cloud," they just called it "the data center." Now, they call it the cloud because it's a shared data center that somebody else manages, rather than a data center in their own building. So, it's nothing new. There's nothing special. I hate to burst people's bubbles, but it's the same thing.

Mala: Do you think it's a big cost saver for a business?

Jason: Oh, no. The cloud is not a cost saver. It's a cost shifter. For instance, I've got clients with five hundred gigs of memory. I've got clients with terabytes of memory. There is no way they would be able to shift that cost to their operations to Azure, or to any cloud platform for that matter, because of resources. When they try to shift that, the cost of a new server would easily be surpassed within just a matter of months. You wouldn't be able to get a server like that in the cloud, for anything cheap. When you're talking about a single server to maintain resources from a high-end on-prem solution—such as terabytes of memory—to a cloud solution, you're talking significant dollar figures each month.

So, what the real magic behind it is, in my opinion, is that you are shifting from a capex expense to an opex expense, and for whatever reason, that model works really well for a lot of people these days. Until they get a year down the road and they say, "Oh, wow. We spent this much?" Yeah, you could have had seventeen new servers for that. It's not a cost saver in many cases.

That said, it offers different or new avenues to be able to provide solutions. Maybe some companies are averse to maintaining their own hardware, and they'd rather have somebody else do it, or they want to be able to scale up or down quickly. That gives you new avenues. For example, if you only have bare-metal hosted servers in your own arena, then you're going to end up with a somewhat restricted ability to scale—up or down—the resources allocated to server A or server B.

Mala: Makes sense. What's your favorite SQL Server feature, if you have one, and why?

Jason: My favorite SQL Server feature, I would have to say, is Profiler. I'm kidding. It's Extended Events, which is much like the cloud. It's still experiencing slow adoption, and people are hesitant to use it unless it has that four-letter word called "Profiler" attached to it. If it has Profiler anywhere in the name, I just don't recommend you use it unless you want severe problems for your server. Extended Events can make your run-of-the-mill DBA look like a rock star, because you have an exceeding amount of information at your fingertips.

I just came back from SQLSaturday in Phoenix. One person who attended a session that I did last year about Extended Events came up and told me how they used what they learned and implemented it in their environment. Subsequently, they were able to resolve problems because of the session that I had given—and their willingness to go and adopt it and utilize it. In this case, it helped them find out how developers were cheating the system and logging in as SA.

Sure, we could use other tools to capture data and events, or to try and figure out which developer is doing what. However, if I were to use a server-side trigger, which is a common practice used to track who is logging in with SA from which machine, I run an extra risk. Server-side triggers can go sideways very easily, and then you can cause all ability to log in to the server to cease, because the server-side trigger is now unable to fire after every connection is made. That's a risk. Extended Events can do the same sort of thing, but it's easier to set up and has a far lower risk related to it. So, it's extremely low risk, extremely efficient, and it's the new way to do things.

Mala: Are there any third-party tools or techniques that you care for and you use?

Jason: I'm willing to go with three third-party tools. One is SQL Prompt by Redgate. If you're not using SQL Prompt and you're using IntelliSense out of SQL Server Management Studio, the question is why? If you have never had a need for code completion, then why have IntelliSense enabled? If you find you like having code completion, SQL Prompt does it far better and with less strain on your SQL Server than IntelliSense does. In the end, just disable IntelliSense and use SQL Prompt instead.

The next one is a tool from SentryOne. When they decided to take Plan Explorer and make it free for everybody—that was huge! Similar to SQL Prompt, if you're not using it, the question is why? The layout and the different tooling that they offer and the features that they allow—it just makes it super easy and super efficient. Sure, you can look at the execution plan inside of Management Studio, and you can look up the XML inside of Management Studio because that's all Plan Explorer is doing—taking that XML and translating it to something that's friendly to you. You get those for free in Management Studio already, you know? Do you need another tool? I would say in this case, yes. You need to have that tool because it's better, especially if it's a really large complex plan. It just becomes easier in Plan Explorer.

That being said, I may not use Plan Explorer every single time I'm looking at an execution plan, because if it's a simple plan, I can open it in Management Studio. It's rather simple. I can identify things quickly. To convert yourself from average DBA to superstar DBA, I'd probably say get that tool.

I've been trying to convince clients to utilize another tool that I like because it can be a cost saver in their environments. I think it's actually been rebranded recently. It was called SQL Clone, and it is now called SQL Provision. Basically, I can take a two-terabyte database and within seconds, I can create a clone of that two-terabyte database.

I'm dead serious when I say "seconds." It takes three seconds to create this clone of a two-terabyte database. On five different servers, six different servers, ten different servers—depending on how many licenses you want to buy for this product. It is a paid product in this case. You decide how many licenses you want to purchase for it, and then you can set up all these clones quickly. So, I can take that two-terabyte database, and I can whip up databases down to the development environments—dev, test, UAT, stage, etc.—or to fifteen developer laptops, if you will, for their dev data versions of the database, and it will be done in just a matter of seconds. I mean it's going to take a bit for that restore to complete even on good hardware, but with this software going to 7800 RPM disk, it takes about three seconds.

Mala: Wow. That is fantastic.

Jason: My two-terabyte database is now—at the completion of the restore process—down to only sixty gigs. That's less than one percent of what it is in production. So that's huge, right? That's a massive footprint that you have removed by saving all of that disk space. Because I'm able to provide that kind of savings, they can have more developers with more individualized development environments. They can each touch different databases, and they now have thirty to forty times the disk space that they used to have. So that's one that I recommend as well because it makes a DBA's life easier to get data down into dev. Oh, and a third benefit, is that it allows you to transform and obfuscate the data so that you can meet your PCI, HIPPA, SOX, GDPR, or whatever compliance you need to check off for the moment. It allows you to alter that data so that it is not actual production data but usable data while you're doing that restore.

It's very cool. It is huge to save yourself that kind of time. For this one client that had the two-terabyte system, it used to take me about 24 hours to get backups restored to them for this two-terabyte database because of the speed of the disk and the resources available to the development environment. That is from the moment they make the request; to me logging in, touching each of the different servers and backups and stuff that I need to do, and then having to make sure that the developers have backed up everything so that we can wipe out their changes. Compare that to the time that same request is submitted to the time that I'm completing it right now, I'm talking a matter of minutes, whereas before we were talking days.

Mala: What sort of struggles have you faced with managers and management, or the business side of things? And how do you recommend handling that?

Jason: Most struggles have been an unwillingness to adopt technology and the changes in technology that were coming, and doing things the hard way for the sake of doing things the hard way.

There are other complications I'm sure every DBA has. It's always the database that's running like crap. The only way you can fix it is to acknowledge what they are saying, and offer them a couple of solutions. Whether they opt to go with your solution or their solution, you just live with it. Don't make a big fuss over it.

Mala: What sort of issues are caused when you interact with other technologists, like SAN admins and network folks, and how do you deal with that?

Jason: The primary issue is tension between the groups caused by poor communication and perceived accusations from one group toward the other. This is resolved when they learn to respect what you can do, and you can learn to respect what they can do, and understand that you can't ever come at them pointing a finger that the problem is their issue. Just like they can't ever come at you and point a finger at you saying it's your issue. You have to come at them and say, "Hey, this is what we are seeing from inside SQL Server. What are you seeing from inside of your metrics that are available at the SAN level or at the OS, because other factors might be at play?" And as soon as you approach it from that kind of a perspective, it disarms them, and then you're able to work together. It is then that everybody really understands that they're working together trying to solve the same situation.

A recent client was dealing with a third-party vendor. Their SAN admins, sysadmins, and admins were looking at an issue and trying to figure out why a third-party product was not communicating with their host server, which was in the cloud, to the on-premise servers. Now, it takes the same exact network path to reach the on-premise servers whether it's development, test, stage, or prod. Everything worked perfectly fine in dev, test, and stage. It just failed in prod. And all the same checks were run on all of the same servers, and everything checked out to be exactly the same on all of the servers. And so, they were looking at it trying to figure out what's going on. The third-party vendor just wrote back a one-line email that said, "Dig deeper!" This kind of response kind of pissed off the SAN admin and network team, and they all had to take a breather about it, because obviously the third-party vendor was blaming them and saying the problem was on their side. But at the same time, it worked for three-quarters of the environment, why was it not working for this one? All you have to come back with is "dig deeper"?

When you come back like that, it's immediately confrontational. Instead, term it differently. Ask if is there is anything else that we could look at. Have we checked this? Have we checked that? This way, you're basically saying dig deeper, but you're not being a jerk about it. Find some different words.

Mala: What's the role that documentation plays in your job?

Jason: Docu what? Does any IT shop really document anything? I'm going to take it from a slightly different perspective because it is part of the documentation. Something that I tell everybody, very consistently, is that they should be using a change control process. If they're not using a change control process, why? The fact of the matter is that change control serves as documentation for what you're doing. How is your environment set up? What has changed recently? When? If you do nothing else to document your environment, but you submit a change control request, and people sign off on it, the worst case is all interested parties are aware of the change that is made, so that's a big win. So, if anything does go haywire, it disarms people from confrontation upfront.

Knowing I can go back and look at change control is a nice comfort in being able to better understand and document that server.

Albeit it's more difficult, I get it, but you have some documentation and you have a history of what's going on with that server, and you know by looking through that change control history who the interested parties are, who the business owners are, the types of applications that are related to that server, and so on, and so forth. You pretty much just start the natural process of documenting it, without ever telling people that they are documenting their environment. When you tell someone they have to sit down and document their environment, guess what they do? People don't want to document it, but if they want to get the stuff changed on their servers, they fill out the change control.

Mala: What are your favorite books, blogs, and things that you follow to keep up?

Jason: Realistically, my most frequented blogs are the Tiger Team blogs. Tiger Team is the SQL Server team. They'll write about different things that they've done. It helps me stay up on stuff that's coming out as well, because they might release something and it may have been very quietly done, but they may actually put something in a blog post about it, and you can discover it there.

I don't do technical book reading anymore unless I am in the mood to study hard. This is in part because I'm browsing random blogs and researching things on a daily basis for a couple hours every day. But it's not even a consistent set of blogs because the blogs that I'm looking through are all technical in nature. I just don't want to sit down with a book to read more technical content unless I really need to study hard.

Outside of Tiger Team and my blog, I'd probably say the one that comes to mind, first and foremost, would be the blog for SQLskills.

Mala: What are some technologies that you are looking to evolve in the data industry? What are you watching with interest?

Jason: Well, every industry is watching virtual reality, right?

I've wanted something in that realm for quite a while, but for a couple of different reasons. One thing that I really want in the virtualization realm is to have Google Glass, or something like that, with the ability to interact with database administration. So that I can just use some sort of VR input into SSMS or servers and be able to manage the databases from that regard without having to bring my laptop and a keyboard, and stuff like that. Just to make things lightweight so I could do it practically anywhere—without having to find a plug-in, or Wi-Fi, or whatever.

The other regard is from the data aggregation perspective, because you know as soon as people start doing VR, somebody is going to want to start capturing all the data, including their precise retinal movements, for their interactions during this VR experience. So, let's go ahead and pull down all the telemetry for the human being so that we can hack them. Though I'm not sure I want to hack certain people, you know. That's the only reason to really grab all that telemetry is so that you can basically steal someone's identity or hack their lives. I'm joking. It will happen, though, as soon as that sort of stuff becomes mainstream and more telemetry is gathered around the VR experience. Just give it a couple years and you'll start finding that people are using VR to hack people's lives. There is positive hacking and there's negative hacking.

So just let me put a positive spin on this. I was watching an NBA basketball game the other night and there happened to be a legally blind child there that had received a VR headset from one of the NBA players. The NBA player paid for the headset and the whole works for this child. The cost for this type of headset is roughly fifty thousand dollars. The only way that this child will truly see an NBA basketball game would be through the use of this VR technology, because it basically gives him the ability to watch the game from within his mind in a virtual reality type of situation, and he can see his favorite NBA players on the court, live as it's happening while he is wearing this VR headset.

It basically hacked his life so he could see an NBA game, whereas he wouldn't have been able to do that before.

Mala: What are your recommendations for stress management and a healthy work/life balance?

Jason: My recommended method for stress management, which I haven't been able to participate in as much as I would like to of late, are actually exercises. I'm pretty sure that most people and most doctors probably agree with that. If you want to manage your stress, your blood pressure, and the stress of IT life, you need to get out and get fresh clean air and exercise. I used to be an avid runner. I've had knee surgery, so I've had to try and find some alternate ways. The way I get around it now is vicariously through my children—I guess cuddling. Or trying to get them to this or that, or it could be an additional type of stress, but at least it's a different type of stress than what work is. And it's a break.

It helps with the work/life balance. It helps change things up. It helps get back into something that's useful.

Mala: Describe your style of interviewing a data professional. What do you look for in people you interview?

Jason: So, I have evolved my style of interviewing for starters. It used to be very technical—I looked for someone who is very competent, technically. And then I evolved. I looked for somebody that has the competency to be technical with a dose of humility. So, I look for them to be able to admit during an interview that they are wrong, because you're not going to be right all the time. I throw in some intentionally controversial questions to see if they will answer with the inaccurate response versus the proven-to-be-true response. To see what they will do and then try to show them the fact of the matter and see if they are willing to acknowledge it. And I found that most are not able to do that. Most people will not admit when they're wrong.

So, if you can't admit when you're wrong in an interview, I don't want you working in the same place as me, because when you screw up a server... well if the person can't admit when they are wrong, and they screw up a server, they're going to deny that they did it. So, you have to set up audits to prove that they did it. It becomes confrontational, and that is not productive. Eventually, I evolved just a little bit more to be more psychological and even less technical. I ask psychological questions to try to figure out if the person will fit into a team type of environment or if they are a lone wolf. I also look for someone who is technically competent.

The tertiary need that I have is somebody who is willing to go out of their way, on their own time, to improve their technical skill sets. I also need somebody that's likely to admit when they're wrong in the workplace. I ask feeler types of questions. I've been pretty accurate so far with those psychological questions.

They have no clue what it is. It's not a stupid "if you're stuck in a blender what do you do first" hypothetical question. They are questions that are real life but can provide me psychological insight into the person.

Mala: What are your contributions to the community? Why do you recommend people be involved with the community?

Jason: I'm pretty active. I do blog semi-regularly right now. I speak at events. I organize events. I offer up free events every now and I'll assist where possible. I also participate in forums. The big reason that I say to do all that—forums, blogging, and speaking—is that if you're putting yourself out there in front of people, where you could potentially be wrong. Some of them have taken what you had suggested and ran your scripts. Sometimes they'll come back and say, "Hey, this didn't work this way in this environment."

So, you are taking a pretty big risk for yourself, and what that does is demand a willingness to adapt and evolve as a professional. To find a way before you put it out there to research even more and to make sure you're getting it right. When you don't get it one hundred percent right, and someone comes along and says you could tweak this and make it better, then be willing to implement that. You will have made yourself a more personable person and a more technically competent person. You have also made yourself more technically competent and marketable in the long run by doing those things.

Key Takeaways

- Be more aggressive in the community arena. Start early.

- The cloud is not a cost saver. It's a cost shifter.

- Change control serves as documentation for what you're doing.

Favorite tools: SQL Prompt, SentryOne Plan Explorer, SQL Provision

Blogs to follow: Tiger Team's blog (blogs.msdn.microsoft.com/sql_server_team/), Kendra Little's SQL Workbooks blog, SQLskills.com

Tim Costello

Data Architect, Brindle Consulting LLC

Tim Costello *is a senior data architect at Brindle Consulting in Dallas, Texas. He has a passion for well-written queries, building fast ETL systems with SQL Server Integration Services and agile data warehouse design.*

Tim wasn't always a database guy. He started out in the woods of northeast Michigan as an EMT (emergency medical technician). He started with the EMS when he was 16 and served in Oscoda, Michigan, for eight years before moving to Texas to finish his paramedic training. Shortly into his paramedic career, Tim took a sabbatical and tried his hand at technology. He started out with Microsoft Access and spent a couple years happily banging away at little reporting systems driven by lots of VBA (Visual Basic for Applications). He taught himself to write SQL queries, design ETL with SSIS and reports with SSRS, and then discovered the joys of visual analytics with Tableau.

The thing Tim is the most excited about today is Biml (Business Intelligence Markup Language). He thinks Biml is the ultimate force multiplier in building ETL solutions with SSIS.

When Tim isn't writing Biml or building a data warehouse, he's probably out riding his bicycle. He aspires to be a long-distance rider. He's gone as far as 75 miles in a day but usually taps out after 40 or 50. One of his goals is to ride a century (100 miles in a day).

Tim blogs at http://timcost.com *and can be found on Twitter at @timcost.*

© Malathi Mahadevan 2018
M. Mahadevan, *Data Professionals at Work*,
https://doi.org/10.1007/978-1-4842-3967-4_11

Mala Mahadevan: Describe your journey into the data profession.

Tim Costello: I started out around 1994-ish, when Windows 95 was out, and it was really exciting and popular. People were hiring anybody that could type to do computer support for Windows 95. I was living in North Texas at the time, and I answered an ad for computer support, it turned out the ad was for Lockheed Martin in Fort Worth, Texas.

It looked like a great job, but I got a little intimidated by the whole idea of moving relocating and all of that. I had never actually been on the Internet as the Internet proper at that time. But I was really active in local dial-up bulletin board systems. And my buddy was an administrator. I was working on his bulletin board, and a lot of other bulletin boards.

So, I called this buddy of mine, and I said, "I've been offered this opportunity. I don't think I'm going to take it, but you should take it." And he ended up jumping on the idea. He moved down to Fort Worth, and he did really well. And about six months later, I had a lot of regret for not following up on that, so I called my friend up, and I said, "Hey, any chance you can get me in out there?" And he said, "Absolutely." So, a couple of days later I had an interview. And about a week later I was loading up a U-Haul truck and moving to Fort Worth.

That was my first IT job. I did second-level desktop IT support, where I would walk around the plant at Lockheed Martin, where they make the fighter jets, and I would actually go to people's desks and replace their network cards and fix their network wiring.

And it was crazy, because they would call the help desk, and they'd say, "Hey, I'm having a problem. I'm trying to do this in Excel." So, they would write down a description of the problem and give it to me, and I would sit down at a computer in their office, which actually was on the Internet, and I would figure out how to do it. I would look it up online, or I would find a book about Excel in our library. I would figure it out really fast. And then I would go to that person's desk and act like I'd always known how to do it.

And I would fix it for them. So, I learned how to figure out problems very quickly on my feet.

My path through the help desk was kind of backward. I started working second-level at the desk. And then I took another job where I was working telephone support. And then I worked my way up through telephone support until I was managing a help desk, essentially.

I hated managing a help desk. I thought I was going to love it. But I hated it because it was all about, "Oh, I don't want to sit next to so-and-so," and "So-and-so keeps burning their popcorn." It was just ridiculous, all of these things that they were coming to me for.

But I found out I really loved working with the data that the help desk generated. Like the daily metrics, the call numbers, the average speed of answers, our SLAs, and all of that stuff. All of these reports were coming to us every day and as the manager, I got to look at them. So, I started spending time with the people that made those reports, and it turned out that they were using Microsoft Access connected to the databases behind the phone switches. They were pulling data from the phone systems. They were putting it into Access databases and creating simple reports and sending them to us.

I thought this was amazing. Soon I started thinking about how they were doing all of this. Back then, I had no idea how to write a query. I couldn't do it to save my life. But I could use Access to drag a table onto the screen, and then another table onto the screen, and then connect two fields. And all of a sudden, I could do all of this cool stuff.

It just gradually became this passion that I began following. I started buying books about databases. I learned about the Kimball approach to data warehouse before I even learned how to write a query.

I was actually trying to write star schemas with many-to-many relationships joined between dimensions and fact tables, and all of this stuff. I was trying to do some relatively sophisticated stuff with Access databases. It was ridiculous, over time, I followed my passion. Access databases led me to [SQL Server] Reporting Services, which brought me to Integration Services. Integration Services brought me to database design, database optimization, and data warehouse design. And, over time, I followed passions around database optimization, or good database design, both third normal form and denormalized. I followed passions about optimizing queries. I tried to learn how to do the administrative stuff, but I was really more of a developer kind of guy. I was really always about business intelligence.

And then I found about data visualizations. I followed a little detour away from Microsoft into Tableau. I spent several years chasing the things that I thought were cool. So, you could say I was following my passions, but I was just having a wonderful time. If I thought an idea, or a concept, or something I saw somebody do was especially interesting or challenging, I would chase that. And it always seemed to fold into the work that I was doing. It kind of kept propelling my career in different directions.

So, at this point, I have a really wide range of experience and a wide range of interests. But at the end of it all, it's still focused on what got me started. That first spark was, "Boy, that data sure is cool. What could we do with it?" And that's my journey from the first day I sat down in front of somebody's computer and tried to help them, to today, when I walk into a conference room and we start talking about how we can use data to make things even better.

It's been a lot of fun. Every day I get to do these things. There's a common element but there's also something surprising and intriguing and exciting just around every corner. And I love it. I mean, almost every day I can't wait to try something new.

Mala: Describe a few things you wish you knew when you started your career that you know now and recommend to somebody who is new.

Tim: Okay, so this is going to sound cliché, but it took me a long time to understand the value of sleep.

In this industry, for whatever reason, we put this premium on the idea of self-sacrifice. "Oh, I pulled an all-nighter." "Oh, I pulled four all-nighters." "Oh, I work sixty hours a week." "Oh, I work seventy hours a week." And it took me a long time in my career to understand that if I focused a little bit more on balance, I could have everything I want. And I would still be able to enjoy it and be much more effective with it.

So, before where I would work eight hours, and then study for four hours, and then worry for four hours, and then do the whole thing over again, now I focus on the work/life balance. I make sure I get enough sleep. I make sure I eat well. I make sure I exercise. And I trust that those things, and my study, and my work will all combine to make me more effective than I would have otherwise been.

I think it's something that's easy to overlook, and I think it's something that's easy to minimize, but it's so much more important than most people realize.

Mala: Describe how a BI professional can expand their skills and embrace the new world of analytics.

Tim: Well, everything is online today. I can remember when it was really difficult to find training online. Now if there's anything that you're even remotely interested in, there are a hundred YouTube videos. And then there are other places where there is higher-quality training that you can purchase. I think it's absolutely, completely justified to go out and purchase high-quality training.

I use Pluralsight. At first, it was difficult for me to pay for the premium membership. But now that's an expense that I don't even consider anymore because I've benefited so much from being able to get out there and do a deep dive into any topic that I want.

So, I think it's important to take a little bit of time, do a self-inventory of the things that are exciting to you, and actually follow your passion to learn about those things, rather than look at the requirements for a particular job, and then learn just enough about that job to get it done, and then move on to the next job.

That's where I started. When I was at the help desk, I would pick up a book about Excel, read three pages, and that's all I cared about. Now, it's a little different. Now, I care about focusing my attention, time, and energy, on things that I'm excited about, and my enthusiasm carries me much deeper into those topics. And I learn so much more.

So, I prefer to go a little deeper into a fewer number of things, rather than just an inch and a half deep on a thousand things.

Mala: So, you recommend that people have to invest in training, learn what they're passionate about, and take it from there?

Tim: Correct.

Mala: What are some ideas one can present to management as a BI person, as far as analytics, big data, and buzzwords like that go?

Tim: I run against the common wisdom in this area. I'm not especially excited about big data or a lot of the buzzwords. There are so many things that most businesses could benefit from that are in the ABC's part of our business. The core, fundamental things that we all learn, we all know, we all acknowledge, but we all ignore.

I think there are a lot of people chasing ideas around big data and data science, and all these buzzwords, when I think the fundamentals are been overlooked. And I think that if we focus on the fundamentals, if we focus on good database design, if we focus on asking the right questions that actually help us work out what the real problem the business is trying to solve is, rather than going out and satisfying this craving to check boxes that the business wants checked that says, "Hey, we're involved in big data. We're involved in data science." I think that's chasing a phantom.

I think it's much better to help people with creating a strong, efficient foundational database system that's well documented, that's well organized, that's consistent, that's easy to understand so that anybody could use it, and any tool can benefit from it. I think that gives you so much more benefit than spending a year trying to chase down fifteen technologies to try and create this complex stack that brings data from the Cloud, through a whole bunch of levels, and then back into the same report that you could create if you spent three months just doing things correctly.

Mala: So, get the basics right.

Tim: Yep. I think it's all about the basics. I think the buzzwords are sexy and they're exciting, but I think that if people focus more on the basics and just sprinkle in a little bit of that other stuff, I think that we would move further as an industry.

Mala: What's your experience with agile and business intelligence?

Tim: That's an interesting question. I never had anything to do with agile in the database world until about two or three years ago. But I've been exposed to it for about ten or fifteen years in the more general development world.

One of the things that have always been important to me is to expose myself to developer groups and to see what people are excited about outside of my primary focus. So, I do go to database groups, but I started, before I ever went to a SQL group, I went through three years of dot-net groups, C Sharp groups, VB dot-net groups, ASP dot-net groups, and user experience groups, and every group I could find. I would go to a group meeting two or three times a week. And inevitably they would all talk about agile.

But I never saw agile in the database world until about two or three years ago. And I find that agile works really well in the database world. The idea of two-week sprints, the idea of taking time to focus on what things are important in the next two weeks, prioritize the things that we can actually get done in a sprint, put time and thought and effort into planning our work.

And then the two weeks sprints keep us accountable. The daily stand-ups keep us focused. Now, I like kanban to help me visualize what my sprint is looking like. But that's just sort of an embellishment. I think at the end of the day the most valuable thing about agile is that it forces you to actually sit down and think about what you're about to do before you do it.

I've found that so many of us in the database world have this luxury of being able to just sit down and bang things out until we get it right. And sometimes we could chase our tail for six weeks to get something that actually works. Where, a sprint might force us to put more thought into it. Very few people actually sit down and think about what they're going to do. They just sit down and try things until it works.

Mala: That's very true.

So, have you seen any kind of a specific advantage with productivity, with agile compared to the old waterfall model? Especially with data warehousing projects, because I've heard arguments that go both ways for it. Especially as far as design changes and things like that go.

Tim: Well, I personally feel like agile is perfectly suited for data warehouse projects. I think that the data warehouse projects fold into the agile idea very neatly, as far as a sprint could be designed around a star, and there are certain milestones within a sprint that you could reasonably expect to achieve. And, as you move forward, you can do sprints that bring in new tech and new functionality. Or you could do sprints that reach back and modify things that you've already done.

I'm a practitioner of agile data warehouse design, which is a little bit different from traditional data warehouse design. For the last four or five years, I have focused all my effort into working my design into a sprint methodology, rather than trying to anticipate every question and every answer. Spend two or three months doing due diligence, and then a year and a half producing something, and then finally, inevitably, what you deliver will not be what they thought they were going to get. And I think that's the biggest cause of data warehouse failures.

To get off on a tangent here, there's a couple of ways that you can design a data warehouse. You could look at all of the reports that are currently in use, and basically decompose them and figure out what elements are involved in those reports, and then figure out a way to compose a data warehouse that incorporates those elements. But all you're really doing in that situation is answering the same questions you're already answering, and you're just changing the plumbing in the background, and nobody will know the difference, right? And I think that's a waste of time, and it never really answers questions that people aren't asking yet. And when they start asking new questions, your data warehouse breaks. So, that's a no-go. And I've seen so many data warehouses run around that idea.

The other idea is you start with a database. You ignore the reports, and you focus on what's in the database and to take all of these elements that are in third normal form and then create some kind of a cool data warehouse out of them. But again, that takes a very long time, and at the end of it, you've got this structure that doesn't really answer questions that people are asking. So, it also fails.

But with an agile methodology what you do is you look at business processes. And each business process should be able to fit within a sprint or a defined number of sprints. You design elements of your data warehouse in a dimensional model that answers questions that people are actually asking, and you anticipate questions that they're not asking because you actually understand the business process you actually talk to people. That methodology seems to work very well. I've had a lot of success with it. But it's counter to what we've always done in the past. And that methodology, the one that's successful, the one that folds into the sprints, would never work with a waterfall method because you just can't spend six months asking all of these questions, and then two years building something, and expect it to work.

Mala: One of the much sought-after skills now is data visualization. What's been your approach to this, and how do you recommend learning more about it?

Tim: There are a couple of different tools out there that you can use. Power BI's a very strong tool but I've always been a Tableau guy. The thing I like about Tableau's approach to data visualization is they limit you to a palette of about fifteen or so data visualizations that there because they work. Your toolbox of data visualizations is limited, but the tools that are in that box are all effective. I really like working with the constraints of a smaller number of tools that I know work.

So, for instance, in Tableau, you can't do a Sankey diagram. And there are times when a Sankey diagram is effective. And you can do a Sankey diagram in Power BI. You have to download another control for it, but it's very easy to make that happen.

But my experience has been that, in order to make a Sankey really work, there're a couple of things that have to happen. The developer has to really understand what a Sankey diagram is, how it works, and what story it's telling. The data has to actually work for the visualization that you're using. And, the user has to actually understand what they're seeing and the story it is telling.

A Sankey diagram looks simple on a page, but it's relatively complex. My experience is that very few developers can sit down and actually tell me what a Sankey diagram is supposed to do, and how the data should be optimally staged to make a Sankey diagram work.

But, if we're talking about a bar chart, much different story here. It's a simple visualization. Anybody can look at it and understand how it works. Any data-based person can understand basic ideas around dimensions and measures and proportions. And you can use simple building blocks to tell a complex story, rather than combining complex building blocks that look really cool on the page, but very few people understand.

So, my approach to data visualization is to keep it simple, tell a layered story with your data. You start at a high level, give people an overview of the data. Let people experience the data by interacting with it, so they can zoom in on it, filter it, highlight it, and see the data change as they do these interactions. And then, finally, bring them down into details on demand.

So, if you work with a layered approach like that, and then your data visualizations themselves have layers You work with the proportions of things next to each other, you add in color to tell another story, you add in tooltips to tell another story, you let people drill from one data visualization to another that's more focused on their starting point, that's more focused on what they're investigating. These things are all very intuitive if you keep your visualizations simple and on target.

So, it's kind of like what I was talking about with this idea of data science and Azure and buzzwords? It's easy to get lost in chasing things that are overly complex and look really flashy, but nobody really understands them, nobody really comprehends the message that they're sharing.

Mala: What are some of the new technologies in the BI world that you're trying to keep up with?

Tim: The thing that I'm most excited about right now is Biml. So, it's Business Intelligence Markup Language, it's a subset of XML. It's a programming language that you use to write Integration Services Packages, but it can also be used for Reporting Services or Analysis Services, or even queries in the database. It's actually literally code that writes code. It can be used to automate tedious tasks, and when you bring in dot net constructs, they can create some really interesting results in your packages.

I am very, very excited about Biml. I think it's going to significantly change the game in Integration Services and in ETL overall. I think it's going to bring more of a pattern-based approach to creating ETL systems, which I think is very, very needed. I think things will be much more consistent when people use Biml because one thousand packages will all have certain elements that are all very consistent with each other. It makes it easier to maintain, it makes it easier to read and understand what you're seeing. I just think Biml is a very exciting technology.

So, it's not as flashy as R, Azure, or data lakes, or any of those things. But, to my mind, again it's a simple tool, it's often overlooked. It's just something that sits in your toolbox and people don't see the value of it because it's not as flashy. But I think it's one of the most exciting things in our industry today.

Mala: Now, what do you see the future of SSIS and SSAS given the amount of automation with the data loading and all of that going on?

Tim: I think SSIS has a very strong future. I think Biml is going to help with that, but even without Biml I think it's very easy to create strong, complex packages that have all of the framework that we need as developers to do things like monitor the execution, monitor the errors, log everything, actually move the data so that we have not only the end result that we get the right record and the right table at the right time, but we know all the ins and outs of the path that that data took. So, I'm very excited about Integration Services, and I see a lot of interesting and exciting releases from Microsoft coming out on schedule for making the tool even stronger.

Now, Analysis Services. Here's where I'm going to kind of probably get myself in some trouble, but I don't see very many interesting updates, enhancements, features being added to Analysis Services, with the exception of the tabular model. I personally, and I've been saying this for years, I think Analysis Services in the old model, the OLAP [online analytical processing] model, the cubes …. I think it's going to be deprecated slowly. I think it's dying a slow death right now. And I think that there are much more exciting and interesting ways to get the same result without using an OLAP model.

I think that columnar indexes are very exciting, I think the tabular model is very exciting, I think in-memory techniques to get fast results are very exciting. I think that traditional OLAP where you go through one hundred steps to create a cube, and then to enhance the cube you have to write MDX [Managed DirectX]—that is a path that fewer people are following. And I think that it's going to slowly fade away.

I have never really been excited about Analysis Services. I think it's on its way out. I have no reason to believe that from anything official I've heard, but just my observations are that I don't see a reason to get excited about Analysis Services.

Mala: What's the role that documentation plays in your job?

Tim: Documentation, that's an interesting topic, isn't it? But nobody really does it, and if you asked ten people, you'll get ten different answers about what documentation actually is. I try to write documentation in the code that I write, in the form of comments.

But I have found that it's been ineffective. One of the pieces that I've always written, and I've changed the values over the years, but in every complex piece of code that I've written, I'll always go deep into a stored procedure that usually would get overlooked somewhere, that has a lot of code, and I go into one of the comments and add a little note that said, "If you read this comment, send me an email and I'll send you …." It used to be a dollar, then it was five dollars, and then it's ten dollars. And nobody has ever contacted me because nobody reads comments.

And, the truth is, comments have limited functionality because they might be meaningful to the person that wrote them, but they might not mean anything to the next person that comes along. I've seen a lot of people write comments that describe what they're doing. So, a comment might be, "In the next line of code we're going to choose between this, this, and this." And then they show me a case statement. Well, I could've looked at a case statement and told you that it's this, this, and this. Right?

It's much more effective to say, "We had to make a choice here because if we don't then we're going to have an issue there."

I'm working with a framework that's actually Biml-based, but it could be done in many different ways. But I like code that is a little bit more self-documenting. I'm beginning to shift a lot of my business logic from actual code in stored procedures, or in SSIS packages, into tables. So that I can have a field and a table that actually shows the formula.

So, imagine a transformation where I'm applying some business logic to get a different number. I'm multiplying this by this and dividing it by that. I take that function, or that formula, I put it in a table. And then in another I can add comments that document why I need this calculation. Using Biml I can pull the formula for use in my packages and the table documents the intention of the formula.

So that decouples my logic from my framework, and it lets me change something in one place and have it affect a change in many places. And if I ever want to see what's going on with certain fields, I can look at my control tables and find out what's going on with those fields, rather than having to dig through tens or hundreds or thousands of stored procedures and SSIS packages to find out where I'm creating this new value, and what I'm using to create it.

I'm still working through this. It's something relatively new for me. But my hope is that it's going to get me away from this nightmare scenario where you have a name for a new field or a name for a new value that could have multiple definitions. Or, you have the same definition stored with multiple names. In either case, you can have a lot of inconsistency and a lot of confusion. And, let's face it, the work that I do in BI, it's all about consistency, reliability, and building and maintaining expectations. And as soon as you put out two different definitions for the same value, and people get confused by that, you begin to lose confidence. And if you lose confidence, it's really difficult to get it back.

So, I will spend hours trying to make sure that something is consistent across all implementations, to avoid a situation where someone comes to me and says, "Hey, gross profit is completely different here than it is there. Why?"

Mala: What has been your experience with the importance of communication with the projects that you do?

Tim: I'm sort of a traditional developer here, in that for a long time in my career I really wanted to be just the guy in the basement. I did not want to have a lot of interaction with the clients.

I wanted them to bring me nice requirements, and nice requirement documents, and I just wanted to do exactly what was in the book. You give me a requirement document that says A, B, and C, and I'll give you exactly A, B, and C. And I might know that you don't really mean C. You mean D. But I'm going to give you A, B, and C because that's what you told me you wanted.

Now, I am much more focused on actually talking to the people that are going to benefit from the work that I do. Talking to the consumers of the business intelligence that I helped create. I try to build those relationships early and make them strong so that they can participate to the level that they're comfortable in the design process.

Now, they might not always know that they're participating in the design process, because many times it's just casual conversations, or there's a lot of whiteboarding that goes on, or whatever. But the better I understand them and the better I understand their needs, the better I can actually answer the questions that they're asking, and begin, over time, to interpret that they might be asking for A but I think they're actually looking for C.

So, now I can have this new conversation with them and say, "Hey, I hear you saying this. But based on our previous conversations, is this really what you want because we could actually be looking at this. They're very close to each other. I just want to throw that up there and see how you feel about it." And a lot of times they'll say, "Oh, my gosh. I didn't realize. That's exactly what I need." And then you can cut out this whole six weeks of churn that doesn't help anybody.

Mala: Do you recommend ways of stress management and developing a work/life balance that people should keep in mind?

Tim: Well, it's going to be different for everybody. Right? Some people are going to really benefit from exercise.

I personally enjoy cycling. But a lot of people enjoy running or going to the gym or whatever. So, I encourage those activities.

I spend a lot of time talking to folks that are coming into this field and trying to give them some ideas to maybe avoid some of the pitfalls that I've fallen through. And one of the things that I often recommend is exploring meditation. Meditation's been very effective for me. My meditation is very simple. I still just focus on breathing exercises. Sitting down and doing a ten-minute meditation in the morning and in the afternoon, it just really helps me focus, it helps me center, it helps me calm my nerves and set out with a more peaceful, a more intentional path, than to be in a kind of constant reactive fire-fighting mode where everything is always, exciting, and everything's on fire and spinning around backwards. But that's not really where you want to be all the time. You want to be in a place where you can step aside from that, understand why these things are happening, and start to circumvent some of that excitement and keep everything a little bit more low-key.

So, meditation has done that for me, but for some people, it'll be exercise, for some people it'll be reading. I encourage rituals.

One of my rituals recently has been I actually get up a little earlier, which is very difficult for me, but I sit in a place in my home that's very comfortable and I just enjoy a cup of coffee for ten minutes while I sort of ease into my day. I don't know. When I'm talking to people I try to find out what works for them, and then I encourage them to do those things. Those are simple things that we do for ourselves, and they're the first things that we always feel like we need to sacrifice to get to those fifty-hour or sixty-hour work weeks. Don't do that.

Mala: What's your style of interviewing a data professional? What are some traits you look for? Any examples of questions you ask?

Tim: I love this question. I ask people what they're excited about. I ask people what the last thing they learned because they wanted to learn it was. And I ask people where they go to find new information.

I spend a tremendous amount of time looking at resources online. So, I've very familiar with all of the very mainstream resources. So, if I'm talking to a person that claims they are a SQL expert, and they have been for several years, and I say, "Give me an example of some places that you might go to find an answer to a complex problem" and they can't tell me anyplace that they would go, I know that they're not really serious about this. It's kind of like we all do when we do a Google search for something. There are certain domains that when they come up in your results, you know you're going to get a solid answer from this domain, or from that domain, or from this site, or from this blog, or whatever. You just know those are going to be solid answers.

And if nobody can tell me any of those names, I don't really take them that seriously. So, I look for what they're learning and where they're learning it.

I might ask them a couple of technical questions. I have a couple of go-to questions in my pocket that are very simple. It's those fundamental things that we all know but we don't ever think about. But I don't go too deep into it, and I never ask people to the whiteboard. Ever. I think that's the worst practice.

So, the question that I like to use the most is, "In a query, there's the select, from, and where. What is the having clause?"

What is it and why would we use it? And seven out of ten people can't answer that question. And that shocks me. I can kind of anticipate which ones are going to get weeded out, and it's a very basic question. But I think the basics are very important.

About six months ago, I was doing a round of job searches. So, I did a lot of interviews. And I went to places and people would say okay. They would ask me all the questions and we would talk about all the things. And then they would come out with a list of whiteboard examples, or maybe a bunch of cards, or something like that. And they'll say, "Let's whiteboard this stuff." And I'll say, "Okay. I'll whiteboard concepts with you." I'll whiteboard what a star schema looks like, or what a bridge table looks like. Or I'll whiteboard ideas. But as soon as they ask me to whiteboard a query, my answer is always no. I've been escorted out of more than one interview because I said no. Because I said I won't whiteboard how to draw that query. I've just spent two hours talking everything about databases with you folks. Don't you believe that I can write a query?

And if I were to ever ask somebody to whiteboard something, that's the answer that I would want from them as well.

Mala: What are your contributions to the community, and why do you recommend people be involved with the community?

Tim: Oh, I love the community. My contribution has been strictly as a speaker. I really wish that I had more time to develop to helping organize events, or volunteering at events, but I just don't. I love speaking. So, I volunteer a lot of time speaking at a lot of events.

Recently, I did a couple of user group meetings. I did an in-person meeting. I did a virtual meeting. The most fun thing, and I think the most rewarding thing I've done recently, was a two-day series of talks about the science and history of data visualizations, for middle schoolers at a private school in Dallas. I talked to about a hundred kids over the course of two days, and I gave the same talk, it was a ninety-minute talk, and I gave it six or seven times. It was so fun to engage with these kids. I volunteered the money that would have come from that to support the Girls + Data program that Mindy Curnutt runs because it's a program that I really believe in.

I love speaking and I love encouraging other people to speak. That's how I give in to the community. The things that I take to the community—there's no way I could even list all of the benefits that I've gotten from being an active member of the SQL community. It's such a strong community, but there's such a family feel to it. And I've made deep, meaningful relationships with people all over the world, people that I consider to be close friends, that have helped me with my career, that have given me new ways to look at the things that we all go through as professionals. I have learned a lot about confidence, I've learned a lot about presenting myself as a professional. And I've learned a lot of technical skills from preparing and delivering these talks at different events.

So, the benefit is tremendous. I'm just getting up and talking for sixty minutes.

But the things that I get from it? There's no way to calculate that. It's literally changed my life. It's been probably had the biggest effect on my career or anything that I've ever done.

Mala: What are your favorite books, blogs, and conferences of training? Name a few.

Tim: There are two books that are my favorites right now, and that's *The Biml Book* by Andy Leonard [Apress, 2017] and a bunch of other authors. Then there's also a book by an English gentleman, his name is Lawrence Corr. The book is called *Agile Data Warehouse Design* [DecisionOne Press, 2011]. I think it's a foundational book of how to look at data warehouse design in a new way, and I think it's a game changer. I think both of these books are game changers, and neither one of them is really been recognized for the potential that it has yet. But I think that we will all see it as we move forward. I can see both ideas picking up steam already.

For blogs, I really love the Simple Talk blogs. Consistently, every time that I go to a Simple Talk blog, it's on point and well-written. It has great information. I can trust it. I just know that when I see Simple Talk, it's going to be good.

For in-person training, I've been fortunate enough to attend a couple of Brent Ozar's training courses. And they are extremely well presented. They're very content rich, but the difference between his training and other technical training that I've attended, is he really focuses on providing a great experience for the person receiving the training. So, it's very high energy, it's very high paced, but there's also this human element that he consistently delivers that makes it feel like it's accessible and it's doable.

I'm thinking of Kendra Little, specifically. She's doing her own thing now. But she's a wonderful speaker. She's a wonderful trainer. I love the way that she uses simple drawings in her training to make them more accessible, to bring you and engage you in what she's talking about. I think it's brilliant, and I love what she's doing with that stuff.

The thing that I admire the most about Kendra—and I'm seeing it in other professionals, like Audrey Hammond and a couple of others—is we all have personal lives that are apart from our professional lives. And it comes back to this whole idea that I was talking about earlier where we all sacrifice so much in this chase to be an ideal professional. So, we all lose sleep. We all work extra hours. We all sacrifice time with our families. We all sacrifice time for ourselves.

And it has a weight. And it has an effect. And it has a cost. And I see some people in our community starting to say, "Hey, guess what? I suffer from depression sometimes." "I suffer from depression and anxiety." "I get really angry sometimes." "Hey, my marriage fell apart." Whatever it is, these are things that I see people in our community start to say, "Hey, these things happen to me. I bet one of you out there might be feeling the same way right now. And I just want you to know you're not alone. I am absolutely here for you."

And I've seen Kendra do it. I've seen Audrey do it. I've tried to do it myself. I think it's important. I can't really put my finger on why but I just think that's something I respect.

Mala: Do you have a funny story or an anecdote that you'd like to share to wrap up? It could be anything, it does not have to be work-related.

Tim: There's a story that I want to share that I think defines a lot about who I am and how I have created this career for myself.

So, many years ago, I went to a conference. It was my very first SQL Saturday event, it was in Alpharetta, Georgia. I went up there with another speaker, and this was the only time that I've been to a SQL Saturday since then where I wasn't a speaker. After that, every SQL Saturday I've been to, I've been a speaker. At the end of the day, there was a session at the last session of the day. It was called Under The Radar BI. So I'm a BI guy, and there weren't a lot of BI sessions at the time, so I thought, "Well, that'll be fun."

The guy that was doing the presentation, his name is Dan Murray. He's one of my mentors now. So, he's up there and he's talking, and he had a wonderful slide deck. He had great visuals. He was talking about how to present data in a visual way, which is something that very few people were talking about back then. And he was just basically talking at a high level about what we can do with data, and how we can share it in new ways. Because up to that point, I was still very, very new, and mostly what I was seeing was just query results. I wasn't seeing a lot of charts and graphs and stuff like that.

So, he's doing this talk. And the people are kind of getting into it, and you can kind of see people, they're body language is loosening up, and they're kind of having fun with the jokes and stuff. But then, about fifteen minutes into the talk, Dan says, "Okay. I just need to be honest with you guys here. We're not talking about a Microsoft technology at the moment. We're talking about Tableau."

"Tableau is a tool that I'm really excited about, and it's a tool that you've seen on the screen for the last couple of minutes, and it's something that I want to share with you all." And you could see, as a group, everybody in the room tightened up, arms started to cross, and people started to lean back. It was amazing. Because it wasn't a Microsoft tool. But my natural instinct was to lean forward. I thought, "This is amazing!" I was so on fire for this thing.

So, I followed Dan and another guy from his company to the after-dinner, and I kept buying him drinks and I kept asking him questions. And he finally told me that there was a lot of free training material on the Tableau website. So, I left SQLSaturday on the Sunday after the event, came home and immediately started watching videos. I watched thirty hours of Tableau training video non-stop for three weeks.

I just couldn't get enough. I was so excited about it. But before I even left Georgia, the guy that I was traveling with, I said to him, "You know what? Within a year, I'm going to be working for that guy, with that tool." And I knew nothing about Tableau at that time. I'd never seen a video about it. I'd only seen one sixty-minute presentation. But I thought, "This guy knows what he's doing, and how to do it. I want to learn that from him, and I want to use that tool. And inside of a year, I'm going to make that happen." And then I went home and I started watching the videos and I figured out that this guy actually lives in Atlanta. I thought, "Gee, how am I going to make that happen?" Because I don't want to move to Atlanta.

I started getting out there. I wrote a couple of blog posts. The only thing that I've ever done that went viral was a Tableau visualization that I created, that visualized my resume in an interactive way. And I put that up there, and I shared it, and somebody put it on Reddit, and in a weekend it had five thousand hits. It was amazing.

But I started learning about Tableau. There were no books at the time. I started using it. I started talking to people about it. I started getting really excited about it. And then finally about eight or nine months after that event, I was talking to a friend in Dallas after a SQL Saturday, and I was telling him how excited I was about Tableau, and he went, "I think my company just started using it and we're looking for somebody." So, I went and I got a job at that company, and it was literally five blocks from my house. I could walk there in ten minutes. It was the closest job I ever had.

I took this job. And I am not kidding, two months later, eleven months after I first saw Tableau, Dan Murray calls me up. And I also have to admit that another thing that I did at this time. I started a Tableau user group in Dallas. And it was very popular. We had over one hundred people on the first night, and then over a couple of months, we maintained a pretty good following. We did some really good presentations, I did a lot of the talks myself.

So, eleven months after I met this guy, I get a call from Dan Murray. He said, "Tim, I remember talking to you last year. I've seen the stuff that you're putting on the web. I've seen a couple of videos that you've put up on YouTube. I've heard about your user group. I've heard about your talks. There's a lot of buzz about you at Tableau, and the things that you're doing." He says, "I want you to come work for me."

Mala: Oh, how nice.

Tim: Can you believe that? And I said, "Dan, I want to work for you so badly, but I can't move to Georgia." And he says, "I don't care. I want you to work for me remote, from wherever you are, I just want you to do the work that you do, and I want you to be a part of my team." So, I came home and I talked to my wife about it. She said, "Absolutely not. You just got this wonderful job five blocks from home. There's no way you're going to take a risk on this."

So, I called him up and I said, "Yeah, just can't do it." I said, "Maybe a couple of months, or a year or something I could do it." He said, "Okay, I respect that." And then he calls me back two weeks later. And he says, "Tim, this is me calling you twice. I never call anybody twice." He said, "This is opportunity knocking. You have to answer."

And I took the job, and I helped Interworks, the company that Dan worked for, build a BI team from. I was the fourth person on the team, and when I left there were about 100 people on the team. I helped build a world-class BI team, all because I got excited about something. I leaned forward. I made a commitment to myself. I learned everything I could about it. I said, "I will do this." And somehow, it happened.

And stuff like that happens. It can happen to anybody. All you have to do is believe. And that sounds corny as hell. Absolutely true.

Key Takeaways

- Do a self-inventory of the things that are exciting to you, and actually follow your passion to learn about those things.

- Rather than chase buzzwords, create a strong, efficient foundational database system that's well documented, well organized, consistent, and easy to understand so that anybody could use it and any tool could benefit from it.

- Keep data visualizations simple, tell a layered story with your data. Don't get lost get lost in chasing things that are overly complex and look really flashy, but nobody really understands them.

- Business intelligence is all about consistency, reliability, and building and maintaining expectations. As soon as you put out two different definitions for the same value, people get confused by that, and you begin to lose confidence.

Favorite books: The Biml Book: Business Intelligence and Data Warehouse Automation by Andy Leonard and Scott Currie, Agile Data Warehouse Design by Lawrence Corr

Blogs to follow: Simple Talk (www.red-gate.com/simple-talk), Kendra Little's SQL Workbooks blog, Brent Ozar's blog

Andy Mallon

Product Manager, SentryOne

Andy Mallon *is a Product Manager at SentryOne, SQL Server DBA and Microsoft Data Platform MVP who has managed databases in healthcare, finance, e-commerce and non profit sectors. He is the SentryOne 2016 Community Influencer of the Year, the founder of Boston SQL, co-organizer of SQLSaturday Boston, and a speaker at SQL Server user groups and SQLSaturdays. He is also the founder of Boston SQL, co-organizer of SQL Saturday Boston and a frequent speaker at SQL Server User gropus and SQL Saturdays.*

Andy started working with SQL Server as a support tech for a startup software company. Unlike most "accidental DBAs," Andy knew early on that he loved SQL Server and wanted to become a SQL Server DBA. He shares his passion for SQL Server by blogging at Andy M Mallon – AM² (am2.co) and SQLPerformance.com.

Andy, his husband, and their dogs live in the same community where he grew up, just outside Boston, Massachusetts. He can also be found on twitter at @amtwo.

Mala Mahadevan: Describe your journey into the data profession.

Andy Mallon: I started my very first job at a software company. I worked mostly in tech support. It was a small company, so there was a lot of wearing multiple hats, but it was primarily training and tech support. We did a lot of

© Malathi Mahadevan 2018
M. Mahadevan, *Data Professionals at Work*,
https://doi.org/10.1007/978-1-4842-3967-4_12

database-heavy tech support things, so I learned SQL Server in that first role. I realized that I liked working with databases, and I really wanted to be a DBA when I grew up.

I had a number of tech support roles, increasingly focused on databases, SQL Server, and Oracle, until I finally had the opportunity to have a position as a database engineer, which was effectively a junior DBA job. Then, I got a "real" DBA job, where I was hired to be a full-time official DBA.

Mala: Describe a few things you wish you knew when you started your career and that you know now, and you'd recommend to somebody who's looking to be a DBA.

Andy: I wish that I knew about imposter syndrome, especially since I recognized fairly early that I wanted to be a DBA. I spent a long time thinking that I didn't know enough to be a DBA, and I wasn't quite sure where I would get that knowledge, or how I would do that. Most of the DBAs I worked with early in my career were really fantastic—complete star DBAs that really knew their stuff. I looked at them and thought they knew so much more than me, I wouldn't be able to get a DBA job.

Now, I know that a lot of people end up getting DBA jobs when they're still figuring stuff out. I knew enough to be a junior DBA long before I ever got the actual job. I could have made that jump sooner, but I was too afraid that I wasn't good enough, and that got in the way of me progressing my career faster. I still have those moments where I feel like I'm not good enough, or I don't know something as well as other people do, or I'm somehow faking it in my career.

But now, I do understand imposter syndrome more. I'm able to recognize what it is most of the time and have that internal dialog over whether I'm getting in my own way. Do I actually need to learn something more? Do I need to do more? Am I actually just better than I recognize that I am?

Mala: What's your experience with agile methodologies in regards to database administration?

Andy: I've never worked in a position where we formally used agile methodologies for being a DBA. I've taken some classes for it, and I do find that a lot of the concepts helpful as far as sprint planning and time planning. As a DBA, you can't plan out two weeks' worth of work and stick to that. There are just too many surprises that come up along the way.

But when I studied agile, I fell in love with the idea of taking a chunk of time, working off of a prioritized list of the things that are the most important that need to get done, and picking the stuff that you can do in the set amount of time that you have, even if it's planning one day's worth of work that way. You look at what you have to do for the week and plan a day where you can pick out what you can actually accomplish in the day and be realistic about it. That really helps to get stuff done, and to be able to accurately tell people when you're going to be able to deliver something.

So the area that I feel like I'm using the agile methodology most is with a one-day "sprint", where I plan my day and work through my day as if it were a one-day sprint.

Mala: Describe your experience with cloud adoptions.

Andy: I think an interesting thing with the cloud is that there are a lot of companies that are really excited by the idea of the cloud, but it's hard to take existing software and realize the benefit in the cloud. If you're building something new from scratch, it's easier to build it to work well in the cloud, but if you have existing software, a lot of times cost or other system integrations end up getting in the way of being able to really use the cloud.

Most companies are excited to move to the cloud, but then, in reality, it's a fairly slow move and you have a hybrid data center where a lot of the existing systems, and larger HR and finance systems end up being on-prem because companies can't afford to spend that much money to put it in the cloud.

Then they've got other software that they're able to put in the cloud and slowly start to move it—even though maybe the CTO is all excited about going full-in on the cloud and putting everything in the cloud. They think they're going to save a bunch of money, but in reality, the cloud can be expensive. It's a shift in what you do—a shift in how you spend money and a shift in the type of work that you do, but it's not some sort of magic.

Mala: Terrible for your cost savings.

Andy: Yeah. If you're moving to AWS or Azure, Amazon and Microsoft make money off of that, and the reason they make money is because it's not free.

If it were really saving that much money, Microsoft and Amazon wouldn't be making so much money.

Mala: So, you recommend the cautious approach and understand that it's not going to save you a ton of money up front?

Andy: Yes. Absolutely.

Mala: What are some of your favorite SQL Server features?

Andy: I really love replication. I know that that's an unpopular opinion to really love replication, but for reporting and for getting data from one spot to another, it's so straightforward. It does have its problems. When it breaks and goes wrong, debugging it and digging through it can be hard. The tooling that Microsoft gives you isn't always the easiest to sort through.

But I really love replication for creating reporting environments, for moving data between servers. So it is a favorite.

My other favorite feature is Availability Groups—to set up high availability or disaster recovery environments. The multiple node availability group has the option to scale out reads to other nodes, and it's fairly easily protect your entire database to keep it highly available, or to ship a copy off to some DR [disaster recovery] site.

Again, it's a feature that needs a little babysitting and some expertise to use, but it's a really awesome feature for a DBA to use.

Mala: Are there third-party tools and techniques that you like to use as a DBA?

Andy: Any time I have a new server, I always install sp_whoisactive, and Ola Hallengren's maintenance tools, and the open source First Responder Kit from Brent Ozar Unlimited. Those three tools I install on every server that I administer. I also am completely in love with Plan Explorer and the SentryOne monitoring tools, Plan for performance and query tuning. It offers so much more detail than the regular execution plans in Management Studio. Some days I spend more time in Plan Explorer than I spend in Management Studio.

Mala: Are there any trends in this line of work that you're looking forward to? Any particular technology that you're watching grow?

Andy: It's interesting because technology is always changing, and keeping up with the latest changes is very important to your career because new features and new technology don't usually get adopted right away, but eventually people start using it.

When I was starting my first tech support job, virtualization was a new thing that nobody was using, and now practically everything is running virtual. And now I think the cloud is in that kind of space, where a lot of people aren't using it and don't really see where they'll get their benefit. But there's a lot going on in the cloud. Certainly, Microsoft's investing a lot in their cloud technologies to try to make them compelling and make it easier for people to start using it.

I think everyone knows that the cloud is the next big technology that they need to keep up with, but I also think that from an individual development perspective, what people really need to think about learning and being better at are automation and architecture. When you're in the cloud, you're building things by piecing different services together and figuring out how to connect part A to part B.

People really need to be good at that architecture piece of taking a service like Azure SQL Database, and thinking about how to take all of these services and piece them together to make your big product that you're going to sell and that your company is going to make money off of. There's a lot of automation that goes on with that. A DBA is no longer worrying about where to put data files on drives. It's more about creating a new server and just spinning up another Azure SQL database that is exactly like the one that you did yesterday.

And doing that in the cloud is really easy, but you have to be good at PowerShell and scripting in order to make that happen.

Mala: What sort of struggles have you faced with management on the business side of things, and what's your recommended approach to handling that?

Andy: I think everyone has probably had a job where the business, their management, or other people were resistant to change. I've certainly had my fair share of cases where I wanted to make a change, or I had something that I was advocating for, and I needed to convince other people that it's the right idea. The DBA sometimes gets to just make the decision and tell everyone to follow along, but in most cases, you're not the only person making that decision. You really need to be able to communicate with the other people directly on your team—particularly the people who are going to be making the decisions—that something is the right choice. You need to convince other people that it's the right choice so that everyone is on the same page.

That kind of communication doesn't always come easily for IT pros. We're not always the best communicators. There are certain things in particular— like careful word choice—that people need to think about when they're talking, especially to managers or the business, to make sure that you don't accidentally offend someone.

Sometimes the people who are resistant to change are the people wrote or built the thing that you want to change. If they're a manager and they've been there for a while, you might be changing their "baby." You want to make sure that you don't accidentally put down that thing that you're changing, because although it might not be the right solution now, that doesn't mean that it was a bad choice when they did it. What's "bad code" today might have been the best code to write at the time.

For example, you need to understand that the most important thing when it was written was that it be done that day.

In that case, it doesn't matter how ugly the code is, it had to be done "today." Perhaps it worked great for five years, if that code doesn't perform well now, you need to rewrite it. But people need to steer clear of how they describe that "bad code" because it might have worked great for five or ten years, and that makes it pretty good. It's far too easy for people to accidentally offend someone, and then you can't convince that person that what you're doing is the right choice because you've accidentally offended them by trashing their "baby."

Mala: What sort of issues are caused when you interact with other technologists, like SAN admins and network folks, and how do you handle that?

Andy: In a way, it's usually a combination of the communication that I mentioned before and also everyone is looking out for what's best for them. I'm looking out for my databases and what's best for my databases, and the developers are looking out for their priorities. Oftentimes, maybe what's easiest for me to do on the database involves more work from a developer— and what's easiest for the developer involves more work for me.

Everyone has to be willing to give and take some to figure out what's really best for the company, or what's best for the organization, and sometimes it involves the very uncomfortable conversation of reminding the other people that everyone needs to do what's best for the organization. In the case of the DBA, one of the things that I frequently find myself arguing with developers over is staffing—there is one of me and there are forty of them. They can spread out a little bit extra work across forty people to make my life easier, and sometimes that's the right choice. Purely because there are more developers than there are me, it can be done faster. And in other cases, the right choice is something that is really hard for me, but it's the right choice and I need to figure out how to spend, perhaps eighty hours building something new because it's the right thing, even though I don't want to do it, and I don't know where the time and effort are going to come from.

I've had more than one conversation where I realized that I'm arguing for what's best for my databases, and the storage admin is arguing what's best for storage. And we're both right because we're both arguing what's best for ourselves. You really have to get everyone to come together so that everyone is on the same page with not just what is best for the database, or what is best for storage, but what is best.

Or sometimes it is just coming up with what is best for now.

Mala: What's the role that documentation plays in your job?

Andy: I'm a very firm believer that I shouldn't be the only person that can do anything. Because if I'm the only person that knows something and I have some knowledge locked away in my head, it means that when I go on vacation, if that thing breaks, then they need to call me and ruin my vacation, and I don't like to do that. And it means that if it breaks in the middle of the night, I have to wake up.

So, making sure that things are documented somehow is important. It doesn't have to be the greatest documentation. It just has to be good enough that people can figure it out. For me, that usually means having some comments right in my code that explain what's going on, and then points to something with more information. If I can keep my code self-documenting, it's easy. You keep it up-to-date, because if I change the code, I can change the documentation. If I change a SQL Server Agent job, the description is right there, and I can change my documentation on that job without having to go to a wiki or some other spot.

But when I do have to have some other spot where it's documented, I just make sure that there's a URL, or some pointer, to that other spot from the code so that it's easy for me to find it so I can keep it updated. It's easy for someone else to find it if I'm asleep or on vacation and someone else has to look at it, they can find my documentation easily. If something breaks in the middle of the night, the documentation has to be there so that I don't get woken up, but then it also has to be able to be found. If someone can't find it, it's not very useful documentation.

Mala: What are your favorite books, blogs, and other means of learning?

Andy: For books, one of my favorite books is SQL in a Nutshell by Kevin Kline [O'Reilly Media, 2008]. He gave me an autographed copy after I started blogging, and speaking, and being more involved in the community. That particular copy that Kevin gave me, the gift and the inscription together, mean a lot to me from a career perspective. More importantly, it's a book that I already love—and now I have a treasured copy as well—means a lot to me. It was one of the books I used as I was learning SQL Server because it was really straightforward and I could figure out what I needed to. I was learning on the job and when I needed to figure out how to do something, that book helped. It is definitely my favorite book.

In general, I learn most things through blog posts and through much shorter things. So when new technology comes out, or old technology that I'm using for the first time, I generally go straight to blogs for learning. SQLPerformance. com and SQLskills.com are two of the blogs that I read on a religious basis where when new content comes out. I at least read the title and introduction of every post to see if it interests me because I really love the deep dive content.

The SQLskills blogs come from the whole team of consultants there. They've just got such deep knowledge of the things that they do. They often talk about the internals of what's happening, and that kind of blog post where it describes what's happening on the inside. It helps me really understand how it works, not just what it does.

Mala: Are there any conferences or any in-person learning events that you're particularly fond of?

Andy: My favorite in-person learning events are SQLSaturday and user groups. The local events really give you the opportunity to interact with the class. If you're at a large conference, and you're one of one hundred people in the room, it's a lot harder to interact with the person who's teaching and to get some time one-on-one time with them before or after, or to ask questions during the session. The structure of those larger conferences makes all of that a little bit harder.

When it's a small user group or a SQLSaturday, you can interrupt the session for questions and the time format is usually more open at a user group. It's not a strict one-hour box. There's really a lot of opportunity there to not just have data thrown at you, but to talk to the person and get your questions answered so that you really understand it. You're not just trying to follow along with what's coming at you.

Mala: What are your recommended ways of stress management and developing a healthy work/life balance?

Andy: I love to cook. If I have a stressful day at work, I like to come home and cook dinner. My husband hasn't had to cook dinner in a decade because I love to cook so much.

Mala: He's lucky.

Andy: I'm a lucky man too. He loves that if I have a stressful day, I come home and I cook. And then when I'm done cooking, I've relaxed and I'm no longer grumpy from my stressful day at work. Also, on Friday nights, we do a date night. Every Friday night, as long as we're both in town and not traveling for work, we go out to dinner. We can have that time where we're just talking to each other. It's dedicated personal time where I'm not looking at my phone. Even if I'm on call, we still go out to dinner. I might have my laptop in the car, but it's just making sure that we have some time together.

I also love to go to Maine. Being in Massachusetts, it's a few hours ride to get up to the woods or to the ocean up in Maine. It is someplace that I can go fairly easily in just a couple hours' drive to get away. It is a happy place that I can go to and relax as a great way to take a vacation.

Some companies let you cash-in vacation time and get money instead of taking a vacation. No. Go on vacation. Take a holiday. Relax and get away from work. It's really healthy. When you come back, you're recharged, and it's easier to have those stressful days because you took time off.

Mala: What's your style of interviewing a data professional? What do you look for, and what are some examples of questions you ask?

Andy: When I interview people, I try to avoid the rapid-fire quiz format. I think it makes everyone very anxious. I'm trying to judge the person on whether their answers are right or not, and they're nervous that they're going to get the right answer. Maybe they have a right answer and I don't understand that it's right. I think it makes for a really awkward interview.

What I prefer to do is start off with some open-ended questions about things they've been working on or some general questions about experience working with XYZ technology. If we're using Availability Groups at their job, I might ask that applicant what their experience is with Availability Groups and try to just ask more conversational questions. If I can get them to talk about a project that they worked on, as they mention specific technologies or specific things that they did, I can ask targeted questions to find out how much they know about that or find out how they work.

I think of a lot of things at work as a problem that we're solving. So as they talk about their experience, I can ask, "Oh, you set up an Availability Group to accomplish that. Did it work the first time? What problems did you have?" And then I focus on digging into those specific things that they worked on, and problems that they had, and how they solved that problem. I can dig into that and really get a good idea of what they know, what they've done, and what they're capable of by asking them about those specific things that they have actually done.

The theoretical quiz type of questions are good, and they can tell you something, but I find it's much more valuable, and you find out more if the person tells you about something they actually did. You can have that back and forth with a candidate. Sometimes I might share a story about when I did something and a problem that I had that relates to whatever we're talking about. I can share my problem and then they can tell me how they would fixed it, and we can have a real conversation, like I would with a data professional friend, over the technical things that I run into and give each other ideas. "Did you try this? Oh, why did you do that? Why did you not try it? Did you just not think of it?"

It can be a much more enlightening if they answer, "No, it didn't work, so I just kept trying and eventually it worked. I have no idea why." That tells you something about how deep their knowledge is versus someone that says, "Oh, I looked at the logs and I found this error and had to go back, and I missed a step." It teaches you a lot more from those answers about what they actually did.

Mala: What are your contributions to the community, and why do you recommend people be involved with the community?

Andy: I organize the BostonSQL user group in downtown Boston, and I speak at other user groups. I've helped organize SQLSaturday in Boston, and I speak at other SQLSaturdays, and I blog. I think all of those things are important, in addition to answering forum questions and other community contributions. They're all really important because that's how I learned.

From when I was starting out all the way through now as a more senior person, if I don't know how to do something or I have an obscure error message, I search for it on the internet and hope that someone else has had that problem before. Blogging and other online forums contribute to those search results. Even just sharing problems on a forum by asking questions is as important as answering them, because it gives people something to find so that when the next person has that problem, it's easier for them.

Any way that people can contribute either their problems or their solutions, it helps other people because now when other people have the same problem, they can find someone to talk to about it, they can find an answer, and they can learn.

Mala: So, you were recently awarded the MVP. How has that experience been?

Andy: It has been great. I was awarded last March, so it's been exactly a year. I just got back from my first MVP Summit and it's been really great. In the MVP program, you do get access to confidential NDA information from Microsoft, but that's not the important part—just the recognition that it gives is fantastic. It gives you a little bit of credibility in talking to others, and whether you're job hunting, or just having a conversation, it's an incredible honor to have that credential next to my name that I've received the MVP Award for what I do with the community.

The MVP Award is awarded for your community contributions. I do the community contributions because I love to help people. I'm not getting paid for my blog posts, or my speaking at SQLSaturday and user groups. I do it because I love the community and getting the MVP Award to recognize that just makes me want to continue to contribute to help the community that I love. Also it helps me fight my imposter syndrome by helping me realize that I'm doing a great job.

There are a lot of people in the MVP program that I look up to, and they're my peers now. I still sometimes have a hard time thinking all of the people that I've learned from being my peers now, being at MVP Summit in the same room with all of these really smart folks, has been a really amazing experience.

Mala: I couldn't agree more. I think we've covered everything I wanted to ask. Would you have anything like a funny story to wind up the conversation? It does not have to be work-related.

Andy: Actually, I have a story from a prior employer that I think is quite funny that I just love to share with other data professionals.

They had a development team that was building a really cool new product that they were going to start offering on a new website. That team was allowed to build whatever they wanted, however they wanted—make their own technology choices without having to get the okay from the other stakeholders that might normally be involved. This meant when they were picking their database, they didn't have to ask the database team, and they chose to build it in MongoDB to make it easy on themselves.

They spent tons of time building this new product, and then they launched it and other groups at the company had no idea how to query MongoDB. There were teams that needed to be able to pull data and couldn't because they didn't know how MongoDB worked. So they ended up taking the Mongo database, writing a process to pull data out of MongoDB and put it in SQL Server so everyone could use it internally.

Even after they made this decision about putting everything in MongoDB, they had to have a whole separate project to put an exact copy of everything from MongoDB into SQL Server so that the data could be useful for the company.

I just thought it was hilarious because if they had asked the database team, we would've told them that they're going to be the only people in the company that know how MongoDB works and we're not going to be able to support them very easily. Other teams aren't going to be able to query the data. And it turned out that had they asked us, and had we provided that feedback, we would've saved a whole lot of trouble.

I think that that is a great example of how if you're laser-focused on the problem you are trying to solve, what's best for you might not be the best solution. When you get back and look at it from the whole company's perspective, it was not the best solution, and in hindsight, I think everyone, including that development team, realized that. But at the time, that development team had their blinders on, and the cost was having to create the SQL Server database anyway.

Key Takeaways

- A lot of people end up getting DBA jobs when they're still figuring stuff out. Deal with imposter syndrome early, if you have it.

- A move to the cloud is a shift in where you spend money and is a shift in the type of work that you do, but it's not some sort of magic silver bullet for cost savings.

- Bad code today might have been the best code to write at the time. Sometimes you need to understand that the most important thing when the code was written was that it had to be done that day.

Favorite tools: Adam Machanic's sp_whoisactive, Ola Hallengren's maintenance tools, Brent Ozar's First Responder Kit, SentryOne's Plan Explorer

Recommended conferences/events: SQLSaturday, PASS Summit, Microsoft MVP Global Summit

Blogs to follow: SQLPerformance.com, SQLskills.com

Steph Locke

Data Science Consultant, Locke Data

Steph Locke *heads up a data science consultancy, Locke Data (https://itsalocke.com), and is a Microsoft MVP for her efforts in the data science space. She leads technical communities at local and global levels to help grow people's skills. A seasoned presenter, Steph delivers talks and keynotes at conferences around the world.*

Steph transitioned from a business-heavy role early in her career in business intelligence and then data science. With over a decade of leveraging data to deliver benefit for companies, including finance organizations and startups, Steph has a proven track record of success. An experienced manager and member of leadership teams, she understands how to align data science with organizational goals and integrate it into company processes.

Now Steph runs her growing data science consultancy out of the United Kingdom with remote employees and freelancers distributed across Europe. As well as building her own startup, she blogs, presents, organizes, and codes around data science and DataOps. She's on Twitter at @thestephlocke.

© Malathi Mahadevan 2018
M. Mahadevan, *Data Professionals at Work*,
https://doi.org/10.1007/978-1-4842-3967-4_13

Mala Mahadevan: Describe your journey into the data profession.

Steph Locke: I actually started working at a burger van at a very young age. My job was initially to bake potatoes, flip burgers, and make hot dogs and stuff. And by the time I was sixteen or seventeen, I was managing it. I was looking after eight other young people in two-meter squared of space on most days, so I learned management, multitasking, and workflow optimization at a young age. That probably impacted my career a huge amount. But, the smell of fried onions was permanently ingrained in my skin and my hair. Made me feel hungry wherever I went, but the job didn't really challenge me intellectually.

I stopped working at the burger van, picked up various temp jobs, and before I finished my third year of university, I got a full-time job at Confused.com as a product analyst. I finished my degree at night. So, it was philosophy, lots of reading books and writing essays, which is probably one of the easiest degrees to do from home and without ever having to go to a lecture.

As a product analyst, my job was partly to provide analysis. When I started there was no data. It was from a third-party system. We just had a web browser. And my boss, before I got there, used to send out a daily report on how many people had used the service. He'd sit there with a post-it and do tally tables.

My job was to fix that. I also did a load of other really useful things. Like, work with suppliers, do contract negotiations. I did a load of digital marketing. I really got my hands stuck in a whole load of the business side, which was really useful. It gave me a strong appreciation for what makes money.

I was doing all of that business side of things whilst also jerry-rigging up a pretty badass Excel solution—consuming the website things and building whole amazing racks of reports off the back of it. Building cool forecasting systems to help us work out where we should spend our money. What sort of rate of returns we might get, things like that. But all of that was done in Excel off of some website pages. That wasn't a very good long-term solution, and I could only do stuff with our one product and I couldn't compare across multiple products. So, I convinced somebody to teach me some SQL.

I was a menace to the data warehouse team. I learned to write SQL, accidentally cross-joining and things like that with our ten million customers. So, I was a hazard for a little while, then I got good enough to know when to kill queries when I had written them badly. By that point, the data warehouse people and the BI team realized that it might be worth bringing me in from the cold and actually keeping an eye on me.

I moved into the team. My job was to take all the stuff that I had built in Excel and build a real solution. So, I got to do the database modeling for the data warehouse. Then the ETL in SSIS to get the data in, and then the reporting on top of it to present it back. And then wonderful SSRS reports. Even ones sent by email that looked good on Blackberries and stuff. So, a lot of end-to-end solutions learning experience. Almost everything I have done since has been finding that next thing that adds value and I'm not bored by doing. It's been a fun journey.

Mala: Describe a few things you knew when you started that you know now.

Steph: The biggest thing is do what's important not what's urgent. It's so much better to do the important thing kind of well than do the urgent thing brilliantly. I still would love to be able to focus on the important things better.

Mala: What was analytics and visualization like five years ago? And why is it so hard now?

Steph: So, I actually wrote about this topic in a book called Tribal SQL [Redgate Books] in 2013. So, it is the perfect proof of what I thought things were going to be like. In the intervening years and the things that I said were important at the time, included SharePoint, which is no longer important. But things like understanding how to model your data and database report design. You should have a good grasp of ETL and SQL, and even a bit of HTML and CSS for styling and stuff. And I think almost everything except SharePoint is still pretty good on that list. And also, I had a section where I looked to the future. And the future was SSRS sucks.

Mala: I love that!

Steph: Excel is great but risky. I call out Power Pivot and Power View, which are being rolled into Power BI. Important things that we should look at. Office 365 is going to be where it is at. And then, outside of the stack, I talk about the growth of things. Those two predictions have borne out pretty well. So yeah, I think we've come a long way from the kind of crappy SSRS reports that everybody hated. And we're now in the position where obviously, we still need some SSRS reports, but we've got really nice interactive visualizations and solutions that people can get so much more out of without us having to do things.

And I think a lot of that change has been to do with the interactivity things. Most of that has been down to JavaScript. You know, like D3, Vega-Lite, high charts, and lucid charts. There's just so many data visualization tools out there, and many have been integrated into great tools like Tableau and Power BI that it has that facilitating of people to answer their own questions without having to learn to use an Excel spreadsheet.

Phenomenal. And I think that's why it is so big now. It's because we've got the ability for everybody from the developer to a business person to be able to make things that help other people so much more quickly than we ever had before.

Mala: Describe your experience with cloud adoption.

Steph: The UK didn't have cloud data centers for the longest time. A lot of the blockers to cloud adoption were compliance. Storing data not in the UK, storing it on somebody else's machines, did Microsoft have access to it, if the government's watching us, and you know, that kind of paranoid but legitimate worry sort of things.

Microsoft and the other cloud providers have done a huge amount, in terms of compliance and data protection, and now I'd almost always say that data is safer on a cloud server than somebody's data center.

So, we're seeing a lot more people adopt putting their data in the cloud because they've had a chance to evaluate data protection and compliance issues, and work out where the real risks are.

The other area that I've seen a lot of adoption is kind of that decision loop that I was talking about for a business user. You have almost a technical loop as well. How do we go from needing some functionality, needing to make, add some business value, to getting that business value live? And previously, a big part of that would have been requesting a virtual machine admin, then going through configuration, and then doing the deployment. Everything just took longer. Now people are able to prototype and do things in production much more quickly, because the tooling around cloud computing can greatly reduce that technical loop.

Mala: So, what are the challenges in analytics today? Both from a technology point of view and a people point of view.

Steph: I think we still don't have compliance solved. So, how do we track data through our systems? How do we cleanse data that we don't need? How do we action request the information from a GDPR? How do we protect it for HIPAA and PCDISS?

There are so many breaches that happen on a regular basis, and it's because we still haven't got to the point where we're doing these things well. And we're doing it well across the board. So, that's a big area.

Another big challenge in the data science world now is we're predicting things and saying whether somebody picks a loan, or how much somebody's healthcare bill's going to cost, or whether they're likely to commit a crime. We're making huge assertions about what people will do and then change their ability to access everything from healthcare to the criminal justice system to finance and shopping. We have to be ethical about this. And not just, "the business told me to do it, so I did it." Having orders is not sufficiently a good enough excuse.

And then the final thing is the technical loop and the decision loop. How do we keep shrinking the length of those loops?

We need more and more people to adopt automation and this sort of value-driven attitude because there's still a lot of people whose job is to click Next in GUI. There are still people manually taking backups out there.

There's still people manually making SSRS reports and spending all their time making things align in Excel. There are so many places where we're not using automation.

And, that's the other big challenge to us today. How do we up-skill people, and how do we keep shortening those loop lengths? They're important things.

Mala: Some people think storytelling means reading stories around data, and data should present facts. What's your take on that?

Steph: When you present a fact, you present a story. When we visualize data our job is to help people draw the "right" conclusion as quickly as possible. Our goal is to keep that decision loop small. They don't have the time to dive into the raw data. They need the facts that fit into their mental story as to what they're doing. So, you need a beginning, middle, and end. But, when we're preparing a piece or a dashboard, we're contextualizing that in terms of how it's going to be used. It fits within how somebody operates, and that's their story. I definitely believe that we need to be thinking about how we make this data fit and how to present it so that it minimizes the time to understand it.

Mala: So, the story as a narrative makes sense. Not story as in fiction, right?

Steph: Yes.

Mala: What are the data quality issues that you face when you're dealing with presentation and analytics, and how do you deal with that?

Steph: There is never an end to data quality issues. A common issue is poor front-end validation and if you get garbage in, you get garbage out.

Problems with data modeling. I see a lot of issues arise because the structure that the data is stored in... it makes it difficult for people to get the job done.

So, they might double information in columns or they repurpose something. They do things to get the job done, but the data doesn't play nicely as a result.

A lot of the issues are lack of safety around data and lack of modeling around it. So, in terms of improving things, it's usually a bit more planning and a willingness to refactor. So, if something isn't working, if something takes a hundred lines of code, but it should be two or three lines of code, there should be a willingness to say, "I'm going to spend ninety-eight lines of codes worth of time, making it so that I can write just two lines of code."

If you're starting a new system, it's a great time to do it. But you can also do this piecemeal. Build quality checks in all along the way. The more checks and considerations of verifying that things are right, the better. And assuming you can give somebody feedback and say this thing is wrong, or this thing needs consideration, the better.

Robust reconciliation and strong constraints on the database, like, say something should be two hundred characters and it should be Unicode. We shouldn't just soft truncate. We should give you some feedback. "You've given us too much data, please fix." Things like that.

Mala: So, let's talk about agile.

Steph: In terms of experience with agile and business intelligence, I was in a number of organizations over the years that implemented agile. Usually, scrub methodology and often being top down. "We're going to do agile so everybody we're either going to send you on lots of training, or you're in scrub teams and go!" And I don't think I've ever been in an organization where that's gone well.

I think that top-down decision making is one of the issues. And the other is being forced into a system doesn't help people understand that system very well. Especially with scrum, it ups the meeting workload for a lot of people. They really feel drained by more admin.

So that big kind of issue. And, in BI, people often get used to being able to push out reports immediately. But then somebody comes along and says, "No, you have to do things in the two-week sprint." And then people are asking, "Where's my report?" And you have to say, "It's only going to take me an hour, but you have to wait two weeks for it."

That level of the cycle is too great. So, I think of it in decision loops. Somebody needs to make a decision. They need some data to make that decision. They have to get it, consume it, and then make the decision.

I'm definitely not a fan of agile. But, reducing that decision loop is all important for adding business value. So, for me, it helped me structure our team, our tools, and the way we think about things that come in request-wise. To minimize that decision loop.

Of course, the ideal answer for making that decision loop as small as possible is the information should already be there. And we don't have to do anything. The fastest code is the code you don't run.

Making things available in an easy to consume way for the user is key. So, self-service BI, such as it is, is a great way to go. We set things up and then people can answer questions for themselves. That's the ultimate in speak, but of course, we can't get everything answered.

I'm a big fan of Kanban, so you basically just have a great big to-do list—a doing column and a completed column. And that doing has limits so that you don't spread your focus too much. And then the business people can just go prioritize that to-do list. And they only need to worry about the top things on the list. And it's about getting things, at least for that BAU—you know, the business-as-usual. Getting it out as quickly as you can, in value order. It doesn't have to be a big complicated system. It just has to be tasks on the to-do list effectively.

Mala: What are some new technologies in the big data world that you're particularly excited about?

Steph: I think that we now need to be more willing to put things in different structures than relational, especially because we have more buy in now. Things don't always have to fit on the table for us to do something with it.

So, we've got up for data science along with some systems like H2O and Spark, for doing that at scale. For storing my data in the format that makes it easiest to work with.

When we code a data transfer or an action, or something that needs to happen, we spend most of our time writing the connections, the retry logic, the logging, and the scalability code. We actually spend maybe 20% percent of the time writing actual business logic code.

With Azure Functions, you just write business logic. Everything else is taken care of. And that means you have less wasted time. You're not doing things that don't add value. You're not reinventing the wheel each time. You're just getting stuff done.

And as data professionals, I think we should spend more time thinking in frameworks and systems so that we spend less time doing the same stuff, and spend more time doing the important value-add tasks.

Mala: What's the role of documentation in being a good BI analytics person?

Steph: I think documentation is super important for every person. But I think it's incredibly vital for BI people. And analytics people generally. So, when we're building reports, we're putting in filters and all sorts of things that construct a view of the data. And we expect people to be able to make decisions off of these. They can't make the right decisions unless they understand what's happening. They need to be able to understand that when we say number of sales, we mean things that have not been refunded within a week, that have a salesperson engage them, and all of these other things that would constitute a sale.

We need that to communicate the understanding properly. And that also helps the future you. So, I always try to be nice to Future Me, because Future Me has done a whole load of other stuff and projects. Basically, the knowledge that Present Me has on a system is going to be much better than the knowledge Future Me has. But Future Me is going to have better coding skills. Hopefully.

I need to give Future Me the information about how things worked, and then Future Me will improve things. So, documentation is as much for me as it is for other people. And, one thing that I have consistently found that helps me have better documentation is to have less code.

So instead of having one hundred SSIS packages and needing to document a hundred SSIS packages, I want an ETL system where those one hundred SSIS packages are automatically generated. Then my documentation focus is on the data sources and destinations. Documenting the code that generates those one hundred things. I can do that much more effectively than I can of robustly documenting a hundred SSIS packages.

Mala: What are your favorite books, blogs, and other means of learning?

Steph: Books that have helped me to no end over the years to be generally better include Don't Make Me Think by Steve Crook [New Riders, 2014]. That is a book on designing websites, which sounds like it's irrelevant, but it's all about identifying conventions, when you should use them and when you should break them. This applies to database design, report design, and effective communication strategies. It's helping people adopt things quickly by leveraging conventions, and then knowing when you want to do something unconventional. So, Don't Make Me Think is great.

The Checklist Manifesto [by Atul Gawande, Metropolitan Books, 2009]. It looks at a series of case studies on how checklists have reduced hospital infractions or improved survival rates for mountain rescue. You know the guy who landed the plane on the Hudson River? That was all done basically by a checklist.

So, the aviation industry as a whole thoroughly believes in checklists, and they have effectively a runbook full of these things. In this situation use this checklist and go through it and make sure you don't forget things and end up safely. So, checklists can be super beneficial. Once you have a checklist—a great way of doing things, then that makes it so much easier to automate.

Because you have your business requirements right there.

The Phoenix Project [by Gene Kim, IT Revolution Press, 2018] helps you understand how you can transform the way you run things to add more value. Nudge [by Richard H. Thaler, Penguin Books, 2009] is a book about the default values of things.

Mala: Default values, interesting.

Steph: If you set the default to be, enroll me in a pension, you're more likely to accept the default than change to the nondefault, and as a result, more people are saving toward their future.

They still make the same decision. Do I have a pension or not? But we're unbelievably lazy people who more often than not accept defaults.

So, having a conscious understanding of when we're building stuff, what can we do to produce the best outcome by default, is very helpful.

Mala: What do you do to gain work/life balance or stress relief?

Steph: I'm very bad at this. So, especially since I've started up my consultancy business, that's meant a lot more travel. Of course, I'm running my own business, so now I'm managing people as well and trying to keep the billable rates up and do all the business admin and things. I've settled for less balance on a daily level. More balance on an aggregate level. I might be on the road for five or six weeks quite a bit, and then I'll try and spend at least two or three weeks a month at home, where I do work that doesn't take me here, there, and everywhere. Trying to make every day perfectly balanced is a losing battle.

It is somewhat achievable. And it really helps of course, that I love what I do for a living. It's very stimulating, and other hobbies often pale in comparison to building a new model, or trying out a new package, or reading a blog post. So, I'm very fortunate to enjoy my job.

Mala: What are your contributions to the community, and why do you recommend people be involved with the community?

Steph: I started getting involved in the community when I was twenty-three. I attended a few user groups, did a talk, and then the organizer of the user group was having a baby and moving to a different part of the country at the same time. So, he dropped the user group on me because I was the only enthusiastic one about things in terms of organization. I started running that, and at one point I was running three user groups, looking at a fourth. That's when I realized I was being a bit insane.

So, instead I just made one super group called Microsoft Stack and kept our groups separate. I also now present around the world. I blog. I help people on different stack groups and on Twitter.

It's been phenomenal because I've met so many great people. I have relatively few friends locally. Almost all of my friends live in other countries. So, going to a conference is actually going to see my friends. And they all understand what I do!

When I go to a conference, nobody asks me to fix a printer. So, it's been really great from a social perspective.

And from a continuous development and professional development… I might be listening to ten hours of talks a month, which is one hundred twenty professional development hours a year, plus all the time I spend reading blogs or things like that. Over the past seven years, I've probably spent close to a thousand hours listening to talks and things and learning through that. And so I've gained a huge amount of knowledge from the communities. And it's great to be able to give back. When I do a talk, that might be anywhere from ten people to three hundred people, getting hopefully an hour's worth of useful content.

I'm absorbing one hundred twenty hours a year but I'm giving back something like five thousand hours a year of knowledge, which is a brilliant exchange rate in my view.

Mala: Wow. That's a really fun way to put it. Very practical.

Steph: I'm a data person. I measure these things.

Mala: Is there a funny story you'd like to share? It doesn't have to be work-related.

Steph: I got home from a day of training some people the other day. I'm sitting on the sofa with the laptop doing a load of business work and eating some popcorn, and I managed to knock the popcorn over. One of my dogs is trying to get it from down the side of the sofa, and it's just stressing her out because she can't reach the popcorn. Oz, my husband, very kindly gets the Hoover out to vacuum the side of the sofa, and this freaks out both my dogs, which are both at least forty pounds. And they both climb over me and the laptop, trying to get away from the Hoover. And my Y key goes with them.

So, there I am at seven p.m., the dogs have just crushed me, broken my laptop, and I've got a day of training the next day as well. Why is my life so hard? And then I looked online. Did you know there is a website I found that specializes in laptop key replacement?

Mala: Oh my, no. That would be useful.

Steph: The only problem is my laptop, which blue-screened to death the very first time I turned it on, appears to have a non-standard keyboard for the model. So, there's a permanent reminder of the time when my dogs freaked out and broke my laptop. It was like a week or two ago and I've just been pressing the nubbin from underneath the key to type a Y. I've just been like completely off my game ever since when it comes to typing. Because everything is wrong now. But the shock, the sheer panic from the Hoover turning on was just unbelievably funny. Being stampeded upon, not so much. You know, it's like the cucumber behind cat GIFs. They're very fun to watch.

Key Takeaways

- Do what's important, not what's urgent. It's so much better to do the important well, and then the urgent thing brilliantly.

- We have to be ethical about how we use data for analytics. It is not just that the business told me to do it, so I did it. Having orders is not sufficiently a good enough excuse.

- The fastest code is the code you don't run.

- We should spend more time as data professionals thinking in frameworks and systems like that so that we spend less time doing the same stuff.

- With community, I'm absorbing 120 hours a year but I'm giving back something like 5,000 hours a year of knowledge, which is a brilliant exchange rate.

Recommended books: *Don't Make Me Think* by Steve Crook, *The Checklist Manifesto* by Atul Gawande, *The Phoenix Project* by Gene Kim, *Nudge* by Richard H. Thaler

Jonathan Stewart

Business Intelligence Consultant, SQLLocks LLC

Jonathan Stewart *is a business intelligence consultant specializing in data visualization, data warehousing, and data management technologies based in the United States. An advocate for educating others, he is a public speaker, teacher, and blogger continually teaching people about the Microsoft BI stack. Since 2000, he has been working in the database field with industry leaders in health care, manufacturing, financial, and insurance, and in federal, state, and local governments.*

Jonathan is very active in the community. He has presented on SQL Server, SSIS, SSRS, Power BI, and business intelligence at numerous SQLSaturday events, local user groups, and conferences throughout the United States and around the world. He participates in webcasts, podcasts, and online presentation events. He is also an active board member for his local Columbus, Ohio, PASS chapter.

© Malathi Mahadevan 2018
M. Mahadevan, *Data Professionals at Work*,
https://doi.org/10.1007/978-1-4842-3967-4_14

Jonathan's blog is called SQLLOCKS (https://sqllocks.net). He is a frequent speaker at local, national, and international events and is a board member of the Columbus SQL Server User Group. In his free time, Jonathan works with the Mid-Ohio Regional Planning commission to help improve central Ohio and enhance the quality of life of its residents.

Mala Mahadevan: Describe your journey into the data profession.

Jonathan Stewart: I wrote my first program when I was six, with a TRS-80. So, I got an introduction to the programming/computer world at a young age and liked it from there. But fast-forward to me being twenty-one. I worked on a help desk. I had a phone call that changed my life. I realized that customer service help desk stuff—that wasn't really good for me.

Mala: Yeah, I understand that.

Jonathan: I went to the library and started doing some research on IT careers, because I knew I didn't want to on the help desk. I wanted to be more.

So, I began to research the type of careers and it came down to three choices. I thought, I'm either going to do system administration, network administration, or database administration. I went through the pros and cons. I'm a real big fan of weighing things like that. I ended up choosing database stuff. I remember the next day, I went to work, and I had no idea what a database administrator was or what they did. I didn't know any of that. I said, "Hey, do we have a database administrator?" And the guy laughed, "Yeah. We have one downstairs. His name is Carlos. You should go see him."

So, I went down there on my break, and I met Carlos, who was a really cool guy. I asked him, "What do you do? Would you be willing to show me a couple things?"

I had already done some pre-research, and I had done things online. Remember this is now seventeen or eighteen years ago, so the web wasn't as it is today. But I did enough to learn. I took some SQL because I knew that SQL was the query language for DBAs.

Carlos said, "Yeah, I can show you a couple of things. Whenever you get some free time, come sit with me. I'm free to show you what I can."

He probably thought I was going to come and sit with him, I don't know, once in a while—once in a week for an hour or so. So, I asked him, "What hours are you here?" And he said, "I work eight to five."

Next morning at 7:55, I was there waiting for him. So, every day from eight a.m. to two o'clock, which was my work time, I sat with him. I did that for about seven or eight weeks. I learned SQL stuff to the point where they could offload work to me. I wasn't getting paid or anything like that. I was just learning. Meanwhile, after I was done, I would go to work on the help desk for my eight to nine hours.

So, the IT manager was tremendously impressed. He thought it was the most amazing thing he had seen, for a twenty-one-year-old kid to have this type of initiative. One day, he said, "I've been watching you for the last few weeks. I'm extremely impressed by what you've done. I want to tell you I've created a position for you. I want to make you an offer for junior DBA." My first database job.

Mala: Wow.

Jonathan: That was my intro into the database world. He said, "With your drive and your determination, I know I won't have you long. So, let's go through and start working out a plan of what you want to do long-term and building your career from there."

So, even during my first junior DBA job, I had already started looking at what the mid-level and senior DBA job postings were looking for—to get to know those things, and speak on those things. I told Carlos, "One day I want to be a consultant." And he told me what I needed to learn first.

I'm a consultant now. But some of the things—it wasn't just the technical skills, it's the soft skills too, and I'm still learning those as I go on. But that's how I moved into the data profession.

Mala: Describe a few things you wish you knew when you started your career that you know now.

Jonathan: This is one of the things I was just alluding to. It's not just technical skills that make you who you are. It's the soft skills. If I could go back and tell myself, "Hey, this is cool. You learn technical stuff and read through things." I can blow through things like that. But I didn't really spend a lotta time learning the human side. I'm still catching up, getting better with it, but because I recognize my weaknesses, I'm still getting better at that.

That's one of the things that for newcomers. You can be bright-eyed and know all the technical stuff, but at the end of the day, even if you're dealing with computers and all this advanced analytical stuff, at some point, you learn to deal with people. And that's the thing that I wish that I had known and taken more time to learn, specifically in my younger days.

Mala: A lot of us techies learn the hard way, sadly. So, what was data visualization like even five years ago? What has changed now, and why is it such an in thing now in your opinion?

Jonathan: Well, thinking back five years ago, Microsoft didn't really play in that space. They did some reporting services, but they didn't really care about reporting services. So, it was like a throw-in. Crystal reports were still big, and visualization itself was just something that had experts, like Stephen Few and Edward Tufte.

But it was still in its infancy because most people thought that visualization was something that anybody could do. Throw a pie chart, show some data, and call it a day. Five years ago, while Instagram existed, it didn't exist like it does today. Even Facebook was nowhere what it is today. The big online media thing was YouTube. It was the big data generator.

IBM had a study in 2011 that showed that even back then ninety percent of the world's data had been created. And obviously, with the advent of Snapchat, Instagram, and all these other new technologies, the data is even more prevalent today. So, that's one of the things that's changed. We have a lot more data to show today. One of the things that I teach in some of my classes is that some people think that all this is for business analysts or report writers. It's one of the things, where if you can get in and understand data visualization, you'll be the superstar.

I did a class out in Phoenix a couple of weeks ago. I told them data visualization is going to come to the point where the Oracle DBA was in the eighties, when they were the true superstar. Data visualization itself is not complicated—people's understanding makes it that because they don't understand the power of color or the speed of perceiving things. And I think once high-level managers begin to realize the power of visualization, they're going to want people who have that. There's going to be sought-after positions specifically for it.

Even now I tell people, "You could be a receptionist or you could be the CEO. In the future, everyone's going to need to know how to display their data." We haven't even reached the inflection point yet.

A lot of times when you show critical data, you only have one shot. You show your message to an investor. You may only have one chance to go to that investor. One of the examples of the power of visualization is Twitter. Was Twitter profitable until last quarter? It just became profitable. So, if you think about showing Twitter's finances to investors, how would you show that they were losing money for six years? You couldn't use all red. You couldn't use the negative charts because it looks doom and gloom. So, you have to figure out how to be truthful and ethical while showing the true data, but not freak out your investors because you're losing money.

So, it's that type of key factor of people beginning to say, "Oh. That's an actual skill set." Microsoft jumped into the market with Power BI. In the last year and a half, they've thrown tons of work into Power BI alone.

Mala: What makes a good data visualization? Describe the approach you go through to come up with one.

Jonathan: A book called Storytelling with Data by Cole Nussbaumer. She worked for Google as their data visualization expert. She had actual theories. She talked about using storyboards. I've always called visualization "storytelling."

Mala: Mico Yuk started a workshop at SQL PASS last year. She used storyboarding. It was wonderful.

Jonathan: Regardless of what you're doing, you can have a basic tabular report, or you could have the most advanced dashboard, it still tells some type of story. Cole talks about going through creating a storyboard using sticky notes. I actually use sticky notes too. The first sticky note is the problem. And that's actually profound because one of the problems what we deal with when we're dealing with data visualization we get requests, and we usually see their requests as the problem but the request is not the problem. The request could be someone saying, "Hey, I want to see sales for the last three years." Right, that's not a problem. That's just wanting to see something. But why? What problem are we trying to solve?

So, go from the problem to the who—who the audience it. Is your audience a person? Is it multiple people? That determines your granularity. Then show the how. How are you going to show what you want to that audience and solve the problem. Are you using distribution comparisons? Stuff like that.

Then you decide what type of mechanism. A lot of times we'll get a request, and we want to run to our favorite tool. In the software world, people want to run to Power BI, which is not always the right tool. Sometimes Excel is your tool. Sometimes PowerPoint is the right tool. So, figure out what the right tool is before you design the thing.

And then once you have those four things, then you can figure out your tone. And then once you have all that, then you can decide which graphs and charts you want to put into your visualization. And those six points I got from Cole. They're profound because you can pick and choose and change stuff in and out. They're extremely interactive. You can get good with the process. You can actually do the whole storyboard in less than ten minutes.

And then from there, I build a mock-up. So now you have the mock-up and the story, so the user knows what they're getting before you've built, and you haven't wasted any of your time.

What makes good visualization? The simplest answer is a visualization that accomplishes its purpose. Every visualization won't answer every question. Some ask more questions, and that's okay. Good data visualization is easily understandable.

One of the things that I tell people is that regardless of what the subject matter is, they should be able to understand what's happening in the visualization. Easily perceptible to anybody who reads it. I call that the "sentence test." You should be able to easily sum it up in one sentence. If you have to explain your visualization, it's probably wrong.

Mala: Why?

Jonathan: Because that defeats the visualization. If you got to explain it, you've probably got it wrong because that's the whole point of visualizing is the speed of perception. Just put the data in Excel and show it to them.

Mala: Data visualization requires a combination of analytical and artistic skills, or is it more of one thing or the other?

Jonathan: It's maybe a combination, I guess. I don't think you have to be artistic at all to be good at data visualization. Stephen Few said that data visualization is not the chance for you to show your artistic talents. That's paraphrasing. He said it differently but that was his point. Data visualization is not your opportunity to show that you belong in MOMA or some other art gallery. You can learn everything you need to learn and make good visualizations.

But even the analytical part—you get data from data scientists and business analysts. It's just understanding key principles for how humans read stuff. How humans perceive things. Once you understand those key principles, you can turn around and display that data so that humans can read and perceive it. Good data visualization is the actual implementation of just good principles.

Mala: Some people I've talked to, think storytelling means weaving stories around data, and data should just be present facts. That view is really common nowadays and can be a little confusing.

Jonathan: Storytelling is not so much stories, like a book, but it's your narrative.

You can make a number, which is a fact, say anything you want because of the narrative. People take facts and they can spin them however they want to spin them. It could be facts and still be a lie. You've got to understand a lot of it is the human factor behind it.

Mala: That's interesting. I heard a talk by Kim Reeves on this whole thing. She takes strong objection to the use of that word, story.

Jonathan: I could see why she would take objection to that because of people's perception of it. But it's a narrative. Regardless of what you're trying to show, it's a still a narrative, and a narrative is a story.

Mala: What are some challenges in data visualization today? Technology as well as with the community.

Jonathan: One of the things about data visualization today is the right toolset. You get a lot of tools. Stephen Few actually goes off on this. Tufte too. They both had issues with it because you hear "self-service BI." That's a huge, huge thing. I can't think if it was Few or Tufte, but one of them had a comment about it's not like pumping your own gas. You hear "self-service," and you think it's like pumping your own gas. He said self-service BI will be like mining, drilling for your own oil, refining it, and then pumping your own gas.

Mala: Wow. That is interesting.

Jonathan: But companies sell self-service BI as if anybody can drill for their own oil because it's so simple. And one of the things I've always told clients is that tools don't add intelligence to your data. So, if your data is in an intelligent state, you can do the world with it. But if it's not in an intelligent state, then no tool is going to magically make it intelligent. It doesn't matter if it's coming from Microsoft or Oracle. They can do great jobs of masking problems, but at some point, you're going to have to put it in an intelligent state. So, a big problem with visualization goes back to the source. Is your data in an intelligent state to start with?

I think that part of the problem is that you have these companies trying to sell. They're saying everything they can because they've got to compete against the big sellers, like Microsoft, Tableau, and Click. Those are the big three visualization tools. How do the smaller ones compete? You can do analytics in ten minutes and self-service this and self-service that. So, they sell a dream without telling people that your data needs to be in a format first.

Mala: How do you deal with data quality issues when you're making a visualization? What do you do if most of the data you're asked to work on is bad or is not satisfactory?

Jonathan: Well, one of the things that I talk about in one of my classes is that you have to be careful with the source of your data. Sometimes you have no say over where your data is coming from, but for those cases where you can't be confident in the source of your data, you need to speak up. Do your due diligence. Send some emails or have a nice trail.

Mala: So, you said a little while ago that visualization is an important skill to learn in the data profession. What are some steps a person can take to learn it if they choose to?

Jonathan: For a DBA, if they're comfortable with their IT tools, I suggest to download Power BI. It's free. Play with it. Get used to putting data in a graph or a chart. Then get to a SQLSaturday, or even a user group, and see some people talk about some type of Power BI or visualization. There are tons of very good visualization people out there, even if it's not their specific thing. Even in the SQL world, there are tons of very good visualization people. There's tons of free information and knowledge you can learn from. I learn by doing and asking. Find somebody you admire and just ask. And then when you're doing your own types of things, ask somebody, "How does this look?" See what they say, and take feedback at its true point. If somebody says something sucks, ask them why it sucks.

You don't have to pay lots of money to get into the visualization world. There are tons of free resources all around. There's SQLSaturday all over the US and all over the world. There's always going to be something there. There are all kinds of places to learn stuff.

Mala: What are some of your personal favorite tools and techniques that you use on a daily basis?

Jonathan: My favorite tool is a pen and paper. I do a storyboard, so I actually physically draw. Secondarily, I use Excel. But obviously, my visual tool of choice is Power BI. I'm a Microsoft person. But those are my tools of choice: pen and paper, Excel, Power BI. I can't live without those types of things day in and day out. And technique-wise, it's storyboarding. I storyboard and try to do the one-sentence test to make sure that I'm checking myself. Not fact checking but checking out my work. Always have somebody you trust that you can go to and ask, "Hey, what do you think of this?"

Mala: What role does documentation play in what you do?

Jonathan: One of the great things about having good documentation is when you're doing research, trying to figure out business rules, and processes. Let's say you're doing a measure and DAX in Power BI. If you can look that up and understand not just what the actual formula is but understand the why, that will help you on the visualization side as well.

I've seen some phenomenal documenters. And they'll put in the formula and explain why. Good documentation will tell you why something exists. Why are we reviewing this data? Why are we doing this? Why are we doing that? Now you can understand how to present that to the user because you understand the why. That's where documentation comes into play.

Mala: What are your favorite books, blogs, and other means of learning?

Jonathan: My favorite books are Cole Nussbaumer Knaflic's Storytelling with Data [Wiley, 2015], Mico Yuk's Data Visualization for Dummies [For Dummies, 2014], and Andy Kirk's Data Visualization: The Handbook for Data [SAGE Publications Ltd., 2016]. Those are three books I keep here on my shelf in my office. I read Andy Kirk's Visualising Data website all the time.

Mala: Mico has a website too.

Jonathan: Yes, Mico's too. There are tons of good resources. All you have to do is go on Twitter, and people will flood you with information. Those are some off the top of my head. I learn daily from them.

Mala: What are your recommendations for stress management and developing a healthy work/life balance?

Jonathan: Have multiple phones that you can smash. No, I'm an avid scuba diver. I like water in general. I'm either on it, in it, or under it. I kayak, I swim, and I scuba dive. I live in Ohio. People say, "How do you scuba dive in Ohio?" Well, outside of that big body of water at the top of Ohio, we also have quarries. So, we scuba dive in quarries. Then if I travel somewhere, I try to get some scuba dive into it.

I'm a huge fan of nature. I like to go for walks, especially when I have a problem to solve. I can think when I'm out in nature. I live in downtown Columbus, so I will sit out by the riverfront, and open my paper or write. That's how I keep my balance.

Mala: Describe your style of interviewing a data person. What do you look for, and what are some examples of questions you ask?

Jonathan: I think a good data professional is somebody who is continually curious. No matter what kind of data professional you are, I think curiosity is the key attribute. Whether you're a DBA expert, a report writer, a data scientist—doesn't matter. I think curiosity is the key attribute.

What happens to xyz? I'll ask questions about certain things, trying to figure out how curious a person is. If you don't know the answer to such and such, how far would you go to get it? Will you just stop at Wikipedia, or will you go and try to track down the product manager from Microsoft? How far are you going to go through this process?

Another one I look for is thinking outside the box. So, people who are going to work for me, any of my teams or for my company now, I want to see if you are an outside-the-box thinker? If I present you a problem, are you going to do what's expected, or what's normal? Or are you going to think of a way to get around certain obstacles? Those are the types of things that I look for interviewing people.

Mala: What are your contributions to the community, and why should people be involved with the community?

Jonathan: Well, for me, I do a lot of speaking at events all over the US. Actually, all over the world I wrote a blog post called "Why We Speak," and it's been probably my most popular blog post. But the things that we can do for the community, you can give back, especially because of what we do. You can take somebody who's a junior DBA, and they can go to SQLSaturday, learn something, and within three months become a major. That can increase their pay by ten, twenty, thirty thousand dollars. That's life changing for families.

At the end of the day, the opportunity to change someone's life is what it's all about. And I tell people too that it's a duty. You get to the point that you're an expert, and you have a duty to give that back to people.

Another thing too, not just in terms of giving back to the people, when you speak about stuff, you quickly want to become an expert because you don't want to have someone call you on something you don't really know. You'll never know everything, but it'll make you better by speaking on it. If you want to be an expert something, speak on it because it's going to force you to become better.

I told you my story about how I got started. Had that one person not been willing to let me sit with them every morning, I wouldn't be here today. That's just one person. That one person changed my life, and then obviously his manager gave me a position, but the one person started it. And I tell people all it takes is one person.

I was actually terrified of public speaking. Even if you don't want to speak, and you have no aspirations to speak, there are still so many other things you can do in the community. You can help facilitate stuff. You can moderate stuff. You can help set up SQLSaturday. You can help run SQLSaturday. There are so many different things that need to be done, you can still get involved and help. Even doing things behind the scenes still helps that person down the road learn.

Mala: Are there any conferences or training programs that you're particularly fond of?

Jonathan: Obviously, the PASS Summit. I had a lot of fun at SQLBits in London. I got to speak in Nepal, and it was so eye-opening because it's a country that doesn't have the resources of the United States, right? Getting to see things from their point of view also makes you appreciate a lot more things. So, I had fun there, even though I got E. coli poisoning while I was there. And you've got to attend SQLSaturday. I love SQLSaturday.

Mala: Can you narrate a funny, interesting story to wind up? It does not have to be work-related.

Jonathan: One of my clients would do this visualization report every month. A sixty-five-page report that had everything in it—tabular stuff, pie charts, bar charts, text, paragraphs. The client would send it everywhere every month and so he would dread it. Not just the five days it took him to do it, but the week before, so he was unproductive for like half the month. So I said, "We can automate this and do it in Power BI and save you all this time." He said, "This is how they've always done it."

I eventually got a survey out to see who was using it. He sent it every month he sent it to thirty people. Come to find out, three people were using less than five percent of the whole thing. And he was so distraught when he found this out. Because it was like half his work, right? Are you familiar with the crying Michael Jordan meme? So, one of the jokes about visualization in general is just because you've always done it doesn't mean that's always still the correct way to do it. And so, but what I ended up doing was taking that sixty-five-page report and making it a three-tab Power BI with drill through. And that solved all of the problems.

So, the client got to see that something that was taking him a whole week to do, was solved with a Power BI report with three tabs. That's my somewhat humorous story.

Key Takeaways

- Good data visualization is the actual implementation of good principles. It is not your opportunity to show that you belong in an art gallery.

- Self-service BI will be like mining, drilling for your own oil, refining it, and then pumping your own gas.

- Understanding the *why* will help you with visualization.

Favorite tools: "Old-fashioned" pen and paper, Excel, Power BI

Recommended books: *Storytelling with Data: A Data Visualization Guide for Business Professionals* by Cole Nussbaumer Knaflic, *Data Visualization for Dummies* by Mico Yuk and Stephanie Diamond, *Data Visualization: A Handbook for Data Driven Design* by Andy Kirk

Recommended conferences: PASS Summit, SQLSaturday, SQLBits

Joseph Sack

Principal Program Manager,
Microsoft Corporation

Joseph Sack *is a principal program manager at Microsoft who focuses on query processing for Azure SQL Database and SQL Server. He has worked as a SQL Server professional since 1997 and has supported and developed for SQL Server environments in financial services, IT consulting, manufacturing, retail, and the real estate industry.*

Joe joined Microsoft in 2006 and was a SQL Server premier field engineer for large retail customers in Minneapolis, Minnesota. He was responsible for providing deep SQL Server advisory services, training, troubleshooting, and ongoing solutions guidance. In 2006, Joe earned the Microsoft Certified Master: SQL Server 2005 certification, and in 2008, he earned the Microsoft Certified Master: SQL Server 2008 certification. In 2009, he took over responsibility for the entire SQL Server Microsoft Certified Master program and held that post until 2011.

He left Microsoft in late 2011 to join SQLskills as a principal consultant. There, he co-instructed various training events, and he was a consultant for customer performance tuning engagements. He recorded 13 Pluralsight courses, including SQL

© Malathi Mahadevan 2018
M. Mahadevan, *Data Professionals at Work*,
https://doi.org/10.1007/978-1-4842-3967-4_15

Server: Troubleshooting Query Plan Quality Issues, SQL Server: Transact-SQL Basic Data Retrieval, and SQL Server: Common Query Tuning Problems and Solutions. He returned to Microsoft in 2015.

Over the years, Joe has published and edited several SQL Server books and white papers. His first book was SQL Server 2000 Fast Answers for DBAs and Developers (Apress, 2003). He also started and maintained the T-SQL Recipe series, including SQL Server 2005 T-SQL Recipes (Apress, 2005) and SQL Server 2008 Transact-SQL Recipes (Apress, 2008).

His most popular white papers include "Optimizing Your Query Plans with the SQL Server 2014 Cardinality Estimator" and "AlwaysOn Architecture Guide: Building a High Availability and Disaster Recovery Solution by Using Failover Cluster Instances and Availability Groups." Most recently, he is a writer on the SQL Database Engine Blog. His classic posts can still be found at https://www.sqlskills.com/blogs/joe/ and https://blogs.msdn.microsoft.com/joesack/. His Twitter handle is @JoeSackMSFT.

You can find Joe speaking at most major SQL Server conferences. He spends his time between Minneapolis and Seattle—meaning that he is either cold or wet at any given point in time.

Mala Mahadevan: Describe your journey into the data profession.

Joseph Sack: I got a bachelor's degree in psychology in the early nineties, and so I had no idea I was going to get into this field. I did restaurant work throughout college and then janitorial work. So, I worked at a place called Muffin Man, and I interviewed at a place called IDS which became Ameriprise, which was acquired by American Express.

They hired me for a role called "team clerk." And at first I actually, I almost didn't take it because when I put in my notice at Muffin Man they said they'd give me a two-dollar raise. I think the team clerk was paying like $7.25 and Muffin Man countered, they were going to give me $9.25. I was going to take it, and then I called my mom, and she said, "Oh no. The benefits! The benefits!" And I'm so young at this point I don't care about benefits. But long story short, or long story longer, I ended up taking the IDS job based on the advice of my mom.

And the job was just typing and filing. I wasn't using my psychology degree at all. And again, this is the early nineties, so this would have been 1994 when I started. And by the later nineties, there was a system called SmartData that everybody in my organization used for reporting 401k transactions. So, there would be quarterly reports and monthly reports. And there was one PC on the floor. It was a personal computer that had this SmartData application that actually ran a Sybase backend.

I noticed that there was a language translation option in the tool. I learned that this language was SQL. It seemed more efficient if I could just write the query instead of having to drag and drop everything. And so, I went to the Barnes & Noble in downtown Minneapolis, picked up a generic SQL book, and started teaching myself SQL. I did that for a year where I was getting really good at writing SQL.

By 1997, my boss told me about a new customer relationship management tool called Onyx, which ran on SQL Server. He sent me to a couple weeks of training, including one week in Washington. It wasn't actually at Microsoft. It was at Onyx headquarters.

And so, I learned the CRM tool, and then they sent me to a couple weeks of training for SQL Server 6.5 and 7.0. And that's how it started, and I loved it. I just knew, even from the moment I was doing the Sybase stuff, it just resonated. It was the first time where I thought this is incredibly interesting. I just want more. And it went up from there.

Onyx used a ton of stored procedures. I dove in, and it was terrifying. I remember in the late nineties, there was big demand for this work. I met consultants that would come in and help with the CRM tool. I just remember thinking, "Wow, these people are magic," and then thinking this is something that I want to do.

So that's how I got into SQL Server in 1997. So twenty-plus years of being a SQL Server professional.

Mala: What a journey.

Joe: Yeah. A lot of good luck. And no regrets either. I went to college early and started at sixteen, so by the time I got out I wasn't cooked yet. I didn't know what I wanted to do, and a psychology degree was just kind of arbitrary because it was what I had the most credits for. The idea that I would have landed in this world—no sign of it whatsoever.

Mala: Describe a few things you wish you knew when you started that you know now and you recommend to people starting out.

Joe: I remember my first review when I was a team clerk. And again, it was filing, running things around, picking up reports, and anything and everything related to practical support of big 401k accounts. In my first corporate review, one of the administrators that I worked with said, "Joe does a very good job but he should know that not everything is urgent."

I remember thinking the feedback was interesting, but I didn't actually internalize it. I heard the feedback and I didn't get it. And years later, I totally get it now, which is that you don't have to go full throttle on everything. I'm a little bit conflicted about the advice because I would say when you're starting off, maybe you do need to go full throttle to stand out. If somebody knows they can give something to you, and you treat it with the utmost priority, no matter how big or how small it is, that's a great way to start off, but it's not good for your health to keep doing that for twenty years.

So not everything is a priority, and you can ebb and flow and the world won't fall apart. That's one thing I really wish I had known and internalized a little bit earlier. I wish I had known that because I would often work myself to sickness.

And then I would say as a second thing, there's always an open seat at every table. It's just not always what you would think the seat would be. And by that, I mean, whenever you start a new job, there's always a way to help. There's always some place to get integrated into, but it's almost never what you expected. The reason why I bring this up is I switched across different roles over the years, and I remember there'd be some scenarios where it wasn't the job I signed up for. It wasn't what I was told it would be.

Then I realized that over time, in some cases it's just not a good fit, but in a lot of cases, it's the seat that you have at the moment, and you should sit in it, and you should do a good job. And then often times that leads to other seats opening up at the table. And so, I think that would have been good to know at the very beginning. Just because immediately it's not what you're looking for doesn't mean that it doesn't transform and change over time.

Mala: So as a product manager, you work with a lot of cross-functional teams. That's typically not what a lot of techies are good at. What are some of the things you would like to share as good practices?

Joe: Sure, so first of all, I totally agree that it's true that it's not as common. I will say this: if you can work well across team boundaries, you are golden. You will have a career for years and years. I would say even if somebody feels that they're not good enough at it, I would tell them that it's actually critical that you build those skills and really try to get good at working across boundaries. The people that get ahead or get the opportunities, those are the folks that can talk to all kinds of people. If you have a reputation of building consensus and not being stubborn and egotistical, that reputation gets around.

In terms of cross-team collaboration… Always expect that there will be some pushback across teams. There are some folks who think that at some point there'll be this perfect scenario where everyone always agrees with each other and decisions get made with ease. But this dynamic just isn't very common and you shouldn't expect it.

Mala: What are some of the important things you did to ensure success and the value of the product you managed?

Joe: The product I'm on is SQL Server and Azure SQL Server database. And then the specific team I'm on is query processing. So, I'll talk about it from that perspective. The first thing is, the success, it's absolutely team-based and I'm just one small aspect of that, so I want to make sure that's totally clear. I'm the public facing aspect of it, so sometimes people see the PM and then they give us way too much credit. But we're really just one facet.

Okay, so with that said, you have engineers and architects, and you have people from years ago that envisioned some of the technologies, and these features are just now finally coming to fruition. So, there's this huge iceberg, and you're just seeing the tip of it.

In terms of important things, I would say one area is based on the very start of the life cycle of a specific feature. So, when you're first thinking about a customer pain point scenario, try to look at it from many angles. I think a lot of features that are quite successful are justified and built based on a mosaic of data points and inputs. If we take adaptive query processing introduced in SQL Server 2017, before we built it, we looked at customer support cases. We looked at feedback on Connect UserVoice. We looked at MVP feedback. We looked at a huge back catalog of different blog posts. We looked at architect guidance from North Star documents that were written from the luminaries in the query processing space.

You aggregate this data and eventually this either adds up to something that we should invest in or not. And so, from a product management perspective, I think the key thing is to not take just one or two inputs. Make sure you widen the perspective as much as possible. And of course during this stage you have to be practical. You can't have this infinite processing stage.

So that's based on the beginnings. If we talk about shipping a feature, it's really important you get as much customer input and testing as possible. And with faster ship releases, that's more and more challenging because you have a smaller runway to get that input.

The other thing is make sure that you get hands on with your own features. So as a product manager, you still need to try things out. If an engineer tells you "Hey, XYZ is ready," you should put your hands on it. You should try it out. You shouldn't just hand it over to the customer. You need to have the full experience.

Mala: How do you manage conflicts with stakeholders in the process of managing the features of the product?

Joe: In terms of conflicts, the first thing to remember is that they're going to always happen. Don't overreact. And by the way, these are not things that you learn once. You just have to continually relearn them. Conflict is never pleasant. You just have to remind yourself continually that this is normal. This is actually not in the way of the process, it is the process. So, you're going to get conflict.

The other thing to remember is even if you don't like hearing it, oftentimes that conflict results in a better feature. So, if you're getting pushback from somebody, and you feel that they have a right to an opinion, you should be listening.

And then the third thing—and this one is very much the case in a cloud-based world, we need to use data and customer feedback to drive the way. So, when there's a scenario that we can't get a consensus on—for example, we can't agree on the specific implementation detail or direction, then that's where the customer feedback should really be the north star.

Mala: What's the best part of any feature you've built and why?

Joe: I've been PM [program manager] officially for two years now. During that time, it's been adaptive query processing, so it's really a sample of one. Everything before that was SQL Server consulting. I was a PM for SQL Microsoft Certified Master. So, I have to say adaptive query processing. I am very proud of the query processing team's work. And the next round of features—I'm very excited about that as well.

But any work related to query processing is incredibly rich and interesting. You have to be extremely careful about regressing workloads. You have to think about things from multiple angles. So it takes a whole team of folks with years of experience to really smoke-test each other and reality check.

Mala: What's a difference that you have seen between waterfall and agile methodologies? And do you have any thoughts on either one or the other?

Joe: I think they both could be effective, and then they both could be muddied up. I've worked with different teams that do one or the other. We kind of do a hybrid on our team. It used to be more waterfall. Now it's more agile. What are the ground rules? I'll just roll with it. I'm pretty flexible.

Mala: Describe your experience with cloud adoption among customers.

Joe: I'm part of the Azure SQL database program management team, so it's definitely super exciting to be a program manager at this time because approximately thirty percent of our SQL DB customers in the cloud are running at compatibility level 140, which would unlock, inherently adaptive query processing features. So, to be talking about that so soon after the launch of SQL Server 2017 is pretty amazing.

Imagine if you were a program manager in 2001, and you don't know this yet, but your feature's not going to ship until 2005. Can you imagine that level of required patience? Now we're in the space where something that we started working on seven months ago is going to see the light of day much more quickly.

Working in the cloud allows us to get value out to customers faster. That's the bottom line. And that responsibility means you have to be very disciplined about making sure that what you get out there is robust and tested, and there's this entire release management infrastructure that reinforces that process.

Mala: What are some industry trends with data that you're excited about, outside of what you're working on?

Joe: Artificial intelligence obviously is a huge trend. I would say anybody in data should at least take a look at it to some extent. You have to decide as a professional how much you want to dip your toe into a particular subject, but I think AI [artificial intelligence] is such a rich area. Mala, you've been focused on data science and looking at that. Has that been satisfying to blog about it and learn about it?

Mala: Oh, I've loved it. I wish I could do that for a living.

Joe: Yeah. So, some of these topics if it resonates with you like that, it's just complementary to your existing skill set. I would say this: if you seriously learn about AI, there will be interesting work for you to do over the next few years. It can only help you.

The other thing kind of complementing that is data science in general. Let's say you've been a DBA for twenty years, and you start getting into data science. There is a market for data professionals who can help statisticians and data scientists do their work. And I call them "the bridge." If you can be a bridge between those two worlds, you will have a lot of interesting work to do, and you will have a lot of grateful people on the other side of that bridge.

Some of it—you have to prove it to them. "Hey, the way that you're doing this aggregation using R can be done much faster in T-SQL." If you can give that kind of help, they will appreciate it, and you will have a job.

Mala: Can you tell me about a time you had to persuade someone in authority to follow your suggestion and how you'd approach that?

Joe: I have the most experience in this through consulting because you have to persuade people a lot. Oftentimes, you're doing it in a big corporation where there are people that own their different areas, or they have their turf and they're resistant to any kind of change. So, my number one tool I've used over the recent years is to "show, don't tell" whenever possible. Show. Use data. Use demos. Especially if it's a "religious" debate—somebody's saying, "No, it's not the SAN. It's your query. It's SQL Server." That's probably the classic scenario.

Instead of sitting there and having eighteen debates over email, you instead show the data, show the path, and show the input/output path stats. And be collaborative. "Okay, I believe you. You're most probably right. But let's just take a look. Let's break this problem into smaller units. Could I get your help to do it?" And the same thing for performance improvements. You should be able to say "Okay, I understand. Can you show me a representative workload, and let me show you a before and after?" Show, don't just tell.

Guess what? Every so often when you try to show, you might be wrong. But I would say that would be the biggest thing that I've done to persuade people in authority in particular.

Mala: What are your favorite books, blogs, and other ways of learning? How do you keep up?

Joe: Yeah, how do you keep up? I remember a few years ago, RSS feeds were my friend. And when that became less prevalent… Have you heard of Pocket?

Mala: Yeah, I have. I know Grant Fritchey is a big user of that.

Joe: Oh, yeah. He has it so that it reports off what he reads, which I think is very cool because it's a way of content curating. I don't have it report back to people but I definitely use it. Because I fly so much, if I see something I want to read, I can save it to Pocket for later. And then if there's no wireless on the plane, I've still saved it as a local version of Pocket and then I read it on the plane.

I'm paying attention to Twitter and LinkedIn. Between those two, you'll see different articles flow through and you can just right-click it and save it to Pocket. I used to have a huge list of 120 different blogs that I follow, but I just can't keep up anymore. You have to narrow it down. That doesn't include all the stuff I have to read from internal Microsoft projects.

So, I have it really curated down, where maybe I'll see a few different links pop up over the week, and I'll save them, and even then, it's a challenge to get through it. From a data professional perspective, there's no way to keep up. I love what SQLServerCentral.com does. Is it their Sunday edition or their Saturday edition?

Mala: Database Weekly.

Joe: I love that. I love it when the SQLskills newsletter and the Brent Ozar newsletter do the aggregations.

Mala: What are your favorite ways of stress management and developing a work/life balance?

Joe: Nobody should listen to my advice on this one because I just have not been balanced my whole life. I would say I have not slowed down until maybe five years ago. At my peak, I was consulting during the day and then writing at night.

I would work on the weekends, usually seven hours on a Saturday, maybe six hours on a Sunday. I would work nine hours or so of consulting on Monday through Friday.

I did not have work/life balance until these last five years, and it wasn't because of some big realization. I think it was just because I'm getting older and I'm slowing down. So, I just can't work seventy-hour weeks anymore. It just doesn't work.

I don't judge people who work too much. I love what I do, and so it doesn't make me feel bad that I had to spend a lot of time doing what I loved. I could have probably done a little bit less.

Mala: What you get in terms of that context switching, you lose in terms of focus. It's a trade-off. For everybody.

Joe: Yeah, totally a trade-off. No engineers like distraction and context switching. None of them. Because that work takes concentration and it takes time to get into that mental state to be able to code well and to get into the flow. And so, to your point, it's not good for anyone that they get distracted.

Do you remember life before cell phones? I look at movies in the nineties, where you didn't see people with cell phones and I feel nostalgic because you had these huge moments where you had no distractions. There was nothing right at your side that was like, "Hey, look at your work email. Check your tweet."

Now I think we need concentration more than ever. We need a lack of interruption more than ever. It's never been more difficult to do so because we have everything at our fingertips.

It's tough. Do you want to have that life where it's quick sips of everything and you never go deep and you never feel that challenge of chewing through a problem?

Mala: Right, very much so. There is a guy who does visualization workshops and all that, you might have heard of him.

Joe: I don't, can you tell me more about it?

Mala: He runs a company that does all kinds of fancy data visualization. The best in the world at that. He has a rule that his team should not check email and should not browse anything on the internet when they start the day. Because he thinks that when you start your day like that, it immediately gets you into a very distracted frame of mind, and what you bring to your work is not going to be that good. So, it was a really interesting thought that he shared, that on my team you don't check email in the morning.

Joe: I think that's cool. Now I will tell you this: I mentioned that engineers need that deep concentration, but from a program management perspective, you need to be all over the place, and you have to switch gears all the time. Because I fly from Minneapolis to Seattle almost every other week, I change when I do certain things. So, when I'm writing specifications, or I have to write blog posts or do things that involve deep concentration, I find a way to weave them into my Minneapolis week. And then for high distraction type activities, I keep it for Redmond. I keep it for Seattle. Seattle's all about how many people can I meet up with, how many whiteboard sessions can I have, how many discussions can I get through. I also save that trip for if there are difficult discussions that should really be done in person. I try to save it for that.

When I'm there, I never try to tell myself that I should go into a deep concentration state because it's almost impossible there. But the deep concentration I can push to my weeks in Minneapolis as I have queued up a bunch of stuff that should be totally distraction free. The thing is I can turn off Skype. I can shut down Outlook, and I can write something for two hours and get it done. But even then, two hours, tops. As a PM, you need to be checking all the time.

Mala: Oh, two hours a day of focus is a blessing in today's world.

Joe: Yeah, not bad.

Mala: So, what are your contributions to the community and why do you recommend people be involved in the community?

Joe: I've gone through a lot of different phases when it comes to the community. I would say the area that I'm most proud of personally is when I was the acting program manager for Microsoft certified masters. That was an incredible experience.

And now, for the community, I work more behind the scenes to watch what other community members are doing, particularly from a SQL database perspective. We check behind the scenes to make sure that people are getting answers and that the people who give answers are getting support from us. "Us" being the product management team.

But I would just say this: one thing that makes a SQL Server community so great is the community and the volume of help that our SQL Server community gives is just outrageously high. Robert Davis, who passed away recently, did an epic amount of work for the community. He just answered so many questions. And he did it without reward. That's what was so amazing.

And so, to me, there are a lot of people actually like that who in the SQL Server community and Azure SQL database are really out there helping. If they weren't there, it would be a much different experience. I think SQL Server's success is very much tied to the strong community.

In terms of part of the question where you're like why do you recommend people be involved? To me even if you're an introvert, it's a forcing function for learning. It's also because what's interesting is these questions don't come in tidy packages most of the time, and so you'll get questions that might be missing key pieces of information. It's like a continual crossword puzzle with an unending dictionary and variety of ways that people could potentially ask things. They could also even be kind of mean or rude. So, you have to learn patience while you do it. But it's this continual exercise that's actually good for your career. While you're helping people, you're also helping yourself.

And so that would be the number-one reason. Whenever somebody wants to join the data community, I recommend they start by shadowing and looking at how people are answering questions. And then get hands on. See if you can answer the question behind the scenes. And then finally dip your toe in and start answering questions. Start building up that muscle, because it's the same muscle you use when helping with debugging and development and helping customers as a data professional.

So, in terms of community, it's incredibly important. Microsoft knows that it's incredibly important, by the way. So, there are stakeholders inside the company that they have a very solid understanding of just how critical the community is.

Mala: Do you have a story you would like to share? It does not have to be work related.

Joe: Sure. I have a story, a little one. When I was sixteen, I was in trouble. I was in with the bad crowd. I had no enjoyment of school. I was a sophomore in high school, was being very rebellious. Had a very difficult home life, a lot of strife happening with both parents. And so, my grandfather had the suggestion that I go to boarding school. I saw boarding school as more of a punitive measure, being torn away from my social circle, which by the way was not a bad thing looking back, but at the time was awful to consider.

Their idea was just to get me out of that bad environment. Over the summer, I worked as a dishwasher at a Mexican restaurant called Zarita's in Madison, Wisconsin, so I took a few days off.

My grandfather lived out on the east coast, and we had a few different places lined up to check out as potential boarding schools. And the choices were awful.

By the end of the trip, we realized I was going to have to pick one of the bad choices. I remember being very anxious about taking more time off from work. I didn't want to lose my job at Zarita's.

The morning of my planned departure, my grandfather said to me, "I had a dream last night, and I think you should stay a few more days and keep looking."

I remember thinking, "No way. I can't be late. I have to wash dishes. I'm scheduled Monday." I remember thinking it was the end of the world that I was being asked not to go back to this Mexican restaurant, which no longer exists by the way. I was worried I would make the manager angry. But I did stay a few extra days to hunt for schools. I listened to my grandfather.

We changed my ticket. The next day, I went to a school called Simon's Rock in Great Barrington, Massachusetts. It's a college for sixteen year olds and up. I interviewed, and it was one of those moments in life where it just feels like everything clicks.

And so I started at Simon's Rock as a sixteen year old. I had teachers that I actually listened to and learned from. I learned how to study. I had colleagues that were interested in learning. And my entire world changed.

The reason why I picked this story is because there are times in your life where you may feel that in-the-moment personal responsibilities are so much more important than the big picture. I could have decided to go home early and then go to a boarding school. Things might have worked out—but I don't think the two paths would have been comparable at all.

I go from almost having to redo Sophomore year in high school to all of a sudden being in this incredible school for two years. Now, after two years, I got my associates degree, but money ran out. We couldn't fund it. I didn't have enough to cover things even with loans, and I had to go someplace much less expensive. So I was trying to figure out places to go and I remember just feeling devastated because I loved the teachers. I loved my friends. It was like a paradise school.

I transferred to the University of Minnesota. At the time, things were outrageously cheap. So I transferred there for the last two years to get my bachelor's, but I remember the whole time thinking life is so unfair that I couldn't stay at Simon's Rock for my full four years.

Then in 1992, there was a school shooting at Simon's Rock. Some of my friends were still there, and some of the teachers, of course. But basically, one of the students killed one of our instructors and schoolmates and wounded others.

I remember hearing from my friends who were still at Simon's Rock, and hearing of the trauma that they had from living through this. That kid had basically gone through with an assault rifle. The details are out there. I won't recap them here. But the point is that it left a lot of emotional scars for friends and faculty. And it's such a small school, there were like three hundred people in the school. I remember feeling this tremendous relief that that was not my senior year as would have happened if I had actually been able to pay for the last two years.

I remember feeling terrible, and then I remember feeling grateful. I always kind of use it as a touch point. When I have these periods of friction where life isn't going how I want it to go, I remind myself to be grateful for all the times that I didn't get what I wanted.

Key Takeaways

- You don't have to go full throttle on everything. Prioritize.

- There's always an open seat at every table. It's just not always what you would think the seat would be. Whenever you start a new job, there's always a way to help. There's always a way to get integrated, but it's almost never what you expected.

- Conflict is actually not in the way of the process, it is the process.

Blogs to follow: SQLskills.com, Brent Ozar's newsletter, *Database Weekly*

John Q. Martin

Product Manager, SentryOne

John Q. Martin *is the product manager at SentryOne, looking after SQL Sentry and the core monitoring suite. John is also a Microsoft Data Platform MVP with more than a decade of experience in SQL Server and the Microsoft Data Platform. John is an experienced DBA and developer, and a former Microsoft premier field engineer. Having worked with SQL Server for the last decade, he has gained a broad understanding of how to use, and misuse, SQL Server. John is also EMEA representative on the PASS board. John blogs at blogs.sentryone.com/author/johnmartin and can be found on twitter as @sqldiplomat.*

Mala Mahadevan: How did you get to the data profession?

John Q. Martin: One part of my past that I think is important to call out is that I was not academic or great at school. Simply put, I failed my A-levels. In essence, I was a high school dropout. My route into a career in IT was an accident—a happy one as it transpires.

A long time ago, in a galaxy far, far, away…. Well, around 2003 and in the UK, I started what might be considered a career by working as a technical store man for an IT service provider. I cleaned and maintained hardware, and delivered it between sites around the UK. This was my introduction to IT. From there, I moved over to working as second-line tech support for the

M. Mahadevan, *Data Professionals at Work*,
https://doi.org/10.1007/978-1-4842-3967-4_16

desktop team, supporting end users. It was here that I got my hands dirty by supporting a Windows migration from NT 4.0 to XP.

After a year or so on this, I moved to another desktop support role. This time working with a diverse array of end users. I was also on the periphery of the migration from NT to XP here too. Looking back, these interactions and the time spent working with and supporting our end users helped me develop empathy for those that I support and build systems for today. I want to understand what frustrates them.

My move from general IT to data was again an accident. The application support team needed a new member of staff to take over supporting an estate management application. I foolishly thought, why not? Here I was working with an application delivered via Citrix and hosted on SQL 2000, with SSRS 2000. Because of the limitations of the application, I needed to start getting writing reports and DTS packages to help improve data quality and automate large bulk actions.

At this time, I was blessed to work alongside a man called Martin Henwood. A man that I consider to be my primary mentor when it came to SQL Server and the business world. His technical guidance and sage advice on the intricacies of how businesses work and how I should learn have stuck with me and helped get me where I am today. I was very lucky to have such a generous person who would take the time to help a young oik new to data and IT get a foot on the career ladder. For this, I am indebted to Martin.

My next step was out into the big bad world and working as a business intelligence developer as part of a team implementing a new ERP platform at a manufacturing company. This was an interesting role where I worked again with some great people and I got my introduction to data warehousing on SQL Server thanks to a gentleman called Chris Todman.

Then it was time to go and experience life as a DBA, working as part of a small team of DBAs or as a lone DBA. I was the guy who did the builds and backups, and deployed to live when the development team needed. This was fun. I met some more great people who I still consider friends, and who like me, went on to start actively contributing to the data platform community.

This is where I discovered the importance of effective communication and how it could have a massive impact on the ability of teams to perform their jobs. I still firmly believe that the DevOps culture that we see today is just effective communication between multidiscipline teams who focus on delivering results.

A couple more hops and I found myself at Microsoft as a premier field engineer, working in the same company as Bob Ward, Mark Souza, and others that had helped shape my path and focus on SQL Server technologies. I spent my time as a transactional engineer, where the work would be different from week to week. Starting the week off with a health check for a customer and finishing it off by having to go on-site with a customer to help resolve a critical support ticket that they had opened with the CSS team.

I have to say that Microsoft was an incredible company to work for. I would recommend that everyone who works with the Microsoft Data Platform try to make this happen. From Microsoft, I joined SentryOne—all thanks to what I thought was a throwaway conversation with Nick Harshbarger in the lobby at SQLBits 2015. I joined as a solutions engineer and then made my way to working in product.

My path through my career has taken me to many roles, but it is important to think about what you do outside of work. My community activities have meant that I got the chance to speak at events all over the world, meeting great people and making lifelong friends. If it was not for being involved with my local data platform group all those years ago, I would have never met my amazing wife. I also studied alongside work, gaining a degree from the Open University in the UK via distance learning.

Where my path leads, I'm not sure. There will be twists and turns and opportunities as well as a few setbacks, no doubt. No matter what comes my way, I take the chances that I am given.

Mala: As a product manager, you work with a lot of cross-functional teams. That's typically something a lot of techies are not totally comfortable with. How do you deal with it? What kind of strategy do you use for that?

John: The key thing in my experience is effective communication. So you can be a good communicator, but you can also at the same time be an ineffective communicator. It's about understanding the types of people you need to deal with. So if you look at the likes of developers and things like that, a lot of them are typically given the facts, given the information. Be short, sharp. Give them that and then let them get on with it.

Then the technical product marketing, where they're a little more verbose. They're a little more about building dialog and relationships. It's about understanding the types of personalities that you're dealing with, because you need to be able to communicate technical information quite often in a non-technical way.

It's all about effectively communicating the message to the appropriate people. It's understanding who you're talking to and what's important to them. Do you need to build the dialog so that they trust you and the things you're saying? Or are they quite simply someone who just wants the facts? It takes a little bit of time to understand those people and understand what you're dealing with.

I'm a people watcher. I can sit quite happily in train stations, airports, things like that for hours and just watch the world around me and the interactions that you see. I take that skill into the office. It takes time. Don't ever expect anything to be really quick. Always be prepared to just take a step back and have a look at something. That's how I really understand who I need to speak to. How I need to speak with them to be able to convey that message across different variables.

Here, I deal with our engineering teams. I deal with product marketing. I deal with marketing itself. I deal with the solution engineering team. I deal with sales, as well as having some interaction with the business. So there's a lot of technical and non-technical roles, and quite often they'll be in the same meeting. It's about finding that level, and it takes time. Don't expect it to come quickly, but be prepared to open your mind.

DevOps are something we've been trying to do for a very long time. At its most fundamental level, it is effective communication. That's what's key to a DevOps culture. If your company's moving toward that sort of thing where you're doing DevOps, it will help you when it comes to cross-functional communication.

Mala: What are some significant things to ensure success and value of the product you manage?

John: One of the most important things that we do here at Sentry is something that I've throughout my career is taking on feedback. There are so many great, cool, wonderful, and shiny things that you can do with technology today, but in order for it to be successful, it has to solve problems for people. In order to have something that's successful is to basically canvass opinion and gather as much feedback from either a diverse array of customers in the industry. Find out the pain points, the problems that exist, and then focus on an area. How does this problem affect this particular area? And then that is how we'll decide to understand what you're looking to build.

It's almost like it's released into the wild. It's about establishing and maintaining that constant feedback loop—speaking with people at the technical conferences, and holding regular customer success meetings with key clients. Is the product that I put out something that the people are talking about? Is it making lives easier and better for people? That's one of the things.

We use a framework called the pragmatic marketing framework, which is all about understanding the market problems to help build the right solutions that will help people.

Mala: How do you manage conflicts with stakeholders when you manage the product? Like one person wants this feature, the other person thinks it's going to take too long. That kind of stuff.

John: Yeah, that's a very common scenario. This is where—again, coming back to the pragmatic marketing framework—it's all about understanding. If you've got two or three different people who want different things, it's about saying, okay, for this feature, product, capability—is there a problem to solve in the first place that is impacting people? Have we got anyone that's willing to buy it? Have we done any research to understand what we need to look at trying to deliver? And then that all comes back to building up what we refer to essentially as an opportunity assessment. So there's an opportunity here to build some software, to solve a problem.

And then we'll take that information and we've got a framework essentially for grading them.

It comes back to the point about effective communication. It is making people understand that you need to make decisions with facts, figures, and numbers. You have to be data-driven. Gone are the days of being able to do something on gut instinct. If you're ultimately building something, and it doesn't work, that's a lot of expenditure, time, and effort.

The long and short of it is effective communication between all the key stakeholders. Don't play email tennis. Make sure that everybody's on the same page and that you've communicated effectively. It basically can eliminate conflict very, very quickly.

Mala: What's the best feature you've been part of and why?

John: At SentryOne, working to deliver the new Top SQL tab. More recently, delivering the managed instance capability. We support managed instances. Whilst it's now in public preview, We've delivered that very, very rapidly. But from an agility and responsiveness perspective, I'd have to say managing the SQL Server. I can't pick one, unfortunately.

Mala: What are the differences you see in waterfall and agile methodologies, and why?

John: Waterfall has a place, and I view that as more of a strategic way of managing things. The SQL Sentry solution we have now is nearly fifteen years old. Managing something over that scale of time strategically in an agile methodology is very difficult to do, so having some form of strategic roadmap—you've got to review it. You don't put it in a sentence, leave it for two years, and then go back and wish you'd done something differently six months ago. We review it on a regular basis.

Agile is the delivery and being responsive to small pieces of work that are easily definable and doable within the work timeframe. The difference between being agile and context switching is the difference between being productive and being unproductive. If you can't do context switching, then the teams delivering the work don't know what they're doing when they're doing it, and they start to lose that bigger picture. Whereas with agile—doing it right, they still have that view of the bigger picture, but they're delivering small pieces of work very rapidly.

Mala: Describe your experience with cloud adoption.

John: Cloud adoption is gaining a lot of momentum. Five years ago, I have to admit, I was very skeptical about the way that it was positioned and marketed. The market was way ahead of the capabilities at the time. In that intervening time, I'm happy to say that I was proved wrong.

The rate of development that Microsoft, Amazon, Google, and even Rackspace have put into engineering solutions are in demand.... Azure SQL Database has the ability to run virtual machines in the cloud, basically removing barriers to adoption, both within the enterprise and the start-up community. I'm seeing a lot of interest at SQL conferences.

You don't define how many VMs behind the scenes. It's pay for what you use on a transactional basis. There's an opportunity to save money because that will fluctuate. You don't need necessarily 150 web servers that only get full on Black Friday. You've got the ability to scale up and down as needed, as well as scale out and then scale back in again.

I think there's going to be in the next decade an awful lot of hybrid environments are really going to come to the fore. A lot more utilization of service element platforms. It's going to be building complementary solutions around, so you think about cosmos with Azure SQL database, managed instances, Spark, data bricks, all of those are platform technologies. You're not going to use one for an entire solution, it's about building and compartmentalizing everything in such a way that it's modular and use the right technology for the right solution.

Mala: Right, totally, yeah.

John: That's come a long way, even in the last five years. I think that's going to continue as we go forward. Yeah, like I say, it's something that I've gone from being skeptical about myself, even working for a cloud vendor to firmly believing that it's got a very prominent place in our industry now.

Mala: What are some industry trends that you're really excited about?

John: Containers are becoming very much mainstream. When you look at what containers and text to data platform, they're not necessarily the right fit, because they're typically immutable. You're going after the desired state. I want five replicas, I want three replicas, I want 20 replicas, which doesn't always gel with having a data persistence there that we would be familiar with regard to things like SQL server. I think there's a lot of, as that technology rapidly matures, we can see more and more ways of implementing the data platform, that's something that's going to be a real game-changer, because if you look at the Azure service fabric, which Microsoft recently open sourced, that's another orchestration mechanism, and having spoken to the couple of people, it's really about the needs that you have - you can have some on premises, some on Azure, some in Amazon. You've got a common framework for managing all of it, but typically you want to have a look at something that's stateless so that you can just spin things up and tear them out, whereas something like Azure is more for stateful scenarios at this point in time. Whether that will change, I don't know. Probably.

I think that's something that I do find exciting, as well as the spark or the non-relational data platform technologies, they're coming up as well. Ultimately, there's also going to be a big place for serverless, just because of the potential for things involved with that as well. Those are the three key things that I'm really interested. I can't, like I said, they're all very closely related and building on top of each other, but that's the area that I'm very excited about in the next three to five years.

Mala: How do you measure the success of a product?

John: With regards to our product, there's a couple of different ways that we look at it. Obviously, there are raw sales—the number of customers buying licenses, and the number of licenses those customers are buying. The other key thing that we do is something we refer to as customer success organization. Everyone's familiar with sales. Everyone's familiar with support and customer service. Customer success is all about actually communicating with your customers. For some of the large customers, we may do it every six months. For other customers, once a year.

If you're the only product in the market space and people don't like it, as soon as you get someone else coming into that market space, you end up with a disruption scenario, and they'll jump ship, potentially, because it may be viewed as, it may not necessarily be better, but it may be an alternative they want to try. It's all about obviously selling licenses and building up your customers, but also making sure that those customers you have are happy with the product.

Mala: So true.

John: If they're happy with the product, you can be confident that yes, we will remain with that or continue to grow. Even if others come into the same market as us with disruptive and new offerings, we can be confident that our customers are happy with what we have, and they're willing to communicate with us when we've got a problem, we may have released stuff that has a bug in it, so they'll contact us, we'll get it fixed, and then they could be confident that they could come to us from that perspective as well. We like to make sure that our customers have all this when it comes to the product. We're not building something that we think they need, we're building things that they're actually saying they need, and that's how we look to guarantee success, is because if you've got enough people that are asking for something, they'll be like, well can we do a workaround that helps them get out of that bind? If you've got 20 to 30 people in your customer base asking for that, then that's a clear indication that okay, we need to look at doing this, because that's a large chunk of people that want to be able to do something about it, so let's see if we can service that need. At which point again, you've got people, if they feel you're listening to them, they're willing to engage more.

Mala: Very true. What was the last time you had to persuade someone in authority to follow your suggestion? How did that work?

John: It's quite easy here at Sentry because of the process we spoke about earlier with regards to opportunity assessments and backing up our decisions with facts.

Working collaboratively means that everybody buys into it, and then even if I may disagree with the ultimate decision, I know that we arrived at that decision through a robust and documented mechanism, and therefore I disagree with it ultimately, but I will back it, because that's what we've come up with, and that's the same with a number of my colleagues.

Mala: What are your favorite books, blogs, and other media, and how do you keep up?

John: Conferences, SQLSaturday, user groups—both attending and speaking. Speaking has made me a better learner and helped me understand technology. Europe and the United States have different views on the way things should be done from a data perspective. It helps having that balance.

From a blog's perspective, Denny Cherry and Associates blog is a great source of information for me. Likewise, Paul Randal's Wait Statistics library is something that I can't thank him enough for spending the time to build.

Then obviously there are things like MSSQLTips. And I try to speak with people on Twitter. One thing I did use early on in my career was the SQLServerCentral site. Their Stairways series is very good.

Mala: I'm a big fan of that too.

John: It is particularly great if you actually want more structured learning.

Mala: What are your ways of stress management and developing healthy work/life balance?

John: Good question. I've got a couple of things that make work/life balance a little bit difficult. One is the amount of travel I do.

When it comes to kick back and relax, I play my Xbox. It's a great mind-in-neutral thing. I also ride my motorcycle.

My wife, she's in the industry as well. She just recently joined Microsoft, but she's a consultant, so we do a lot of these things together. We go out for a nice long motorcycle ride in the countryside. We watch motorcycle races—British Superbikes, World Superbikes, and things like that.

Mala: You mentioned a little while ago about speaking and attending a lot of SQLSaturdays. What are some of the reasons you recommend people be involved with the community in general?

John: The community's got a lot when it comes to learning. If you're learning with others, it makes that learning process easier. I've been a lone DBA in a company. If you've not got anyone to bounce ideas off, it becomes a little more difficult, and you feel like there's a lot more weight on your shoulders. The community gives you that pool of people you can bounce ideas off of.

There's a lot of friendships to be made there as well. Otherwise, a blog is a fantastic way of building up your own knowledge base.

Mala: What are some of your favorite tools and techniques with SQL server, what do you use that you're particularly fond of?

John: Plan Explorer. I've used that forever, even before I joined Sentry. RML Utilities. I'm a huge fan of it when it comes to minimizing workloads for performance, but also for migrations, using extended events or trace to capture the workload, and things like that, or just pushing all the metrics into a database to do your analysis on them. You've also got things like Next and PSS Diag, which are a little more obscure when it comes to troubleshooting. Another great resource is Glenn Berry's scripts. Plan Explorer, DBA Tools, RML Utilities, and Glen's scripts are probably my top four.

Mala: What has been your experience with the MVP program? What do you think you've gained out of it?

John: The interaction I have with the product group. The distribution lists and things like that where various discussions go on. It's a lot of useful information. I've already met and spoken with a lot of the people that are on the MVP program. We can have a really good discussion around the technologies that are coming and what's going on.

Mala: Do you have an interesting or funny story you'd like to share? Does not have to be work-related.

John: Yeah. It comes back to what I mentioned about stress management and work/life balance. My wife and I got married in August of 2017. We were looking for honeymoons. I just stumbled across a little advert on Facebook on riding Royal Enfield motorcycles in the Himalayas.

We've got our full bike licenses. Essentially, we flew into Delhi and had a fourteen-hour bus ride. It was one of the most terrifying bus journeys of my life. The roads with the sheer drops. From there, we spent ten days covering just over 1100 kilometers in some of the most incredible and beautiful scenery, riding Royal Enfield Motorcycles. It's a fantastic experience for meeting the people, seeing the scenery, making new friends as part of the tour that we were on.

It was something that pushed us, not from a technical perspective, but getting out there and experiencing something completely different. It was fantastic. I will happily do something similar again, no doubt.

Mala: Thank you, John.

John: Not a problem, cheers Mala.

Key Takeaways

- With communication, don't play email tennis. Make sure that everybody's on the same page and that you've communicated effectively. It can eliminate conflict very quickly.

- You're not going to use one technology for an entire solution. It's about building and compartmentalizing everything in such a way that it's modular. Use the right technology for the right solution.

- We're not building something that we think they need. We're building things that they're actually saying they need, and that's how we look to guarantee success.

Favorite tools: Sentry Plan Explorer, DBA Tools, RML Utilities, Glenn Berry's scripts

Recommended conferences: PASS Summit, SQLBits, SQLSaturday

Blogs to follow: Denny Cherry & Associates Consulting blog, SQLskills.com, MSSQLTips.com

John Morehouse

Consultant, Denny Cherry & Associates Consulting

John Morehouse *started his IT career many years ago at Mutual of Omaha in Nebraska. He eventually became an accidental DBA and worked with SQL Server full time in 2007 as a consultant for Sogeti. Later, he became a database architect for Farm Credit Services of America. During his time in Nebraska, he rebooted the Omaha SQL Server User Group and led the group until late 2014, building a distribution list of more than 800 members.*

In early 2015, John made the move to Farm Credit Mid-America in Louisville, Kentucky. Currently, he works for Denny Cherry & Associates Consulting, helping clients solve complex technical issues with SQL Server. In 2016, he took over leadership of the Louisville SQL Server User Group and continues to lead it today.

John has helped organize numerous SQLSaturday events. He is a PASS Summit speaker and is heavily involved with the PASS organization. He is a two-time Outstanding PASS Volunteer award winner, a Microsoft Data Platform MVP, a Friend of Redgate, a SentryOne PAC member and Community Ambassador, and a 2016 Idera ACE. John firmly believes in giving back to the #sqlserver community.

John is an avid blogger and enjoys reading novels. He is also the Cubmaster for his son's Cub Scout pack and greatly enjoys spending time with his two sons. John also enjoys traveling and seeing new places.

You can find John online at www.SQLRUS.com or on Twitter as @sqlrus.

© Malathi Mahadevan 2018

M. Mahadevan, *Data Professionals at Work*,
https://doi.org/10.1007/978-1-4842-3967-4_17

Mala Mahadevan: Describe your journey to the data profession.

John Morehouse: My journey into the data profession is your typical accidental DBA. Years ago, I worked for a company in Omaha, Nebraska. Really, I was a jack-of-all-trades. I was specializing in Lotus Notes. I was on the internal IT staff, and so not only did I do Lotus Notes administration and development, I was also the webmaster and basic IT guy. We had SQL Server in house, but my counterpart is the one that actually managed it and was responsible for it. He left for a different position, and I went to my boss and said hey, I want to be the SQL person, if you will. I actually took a hard look at what I thought was going to have longevity in the market, and I figured that data in some capacity would always be a need. This was back when the internet dot-coms were exploding and coming up, so I figured data would always be a profession that would need people. I wanted to learn SQL Server and really focus on that. I took that on for a little bit, and was the accidental DBA learning as I go. I will admit that I had a bunch of cursors that I thought were fantastic, and now I know better.

In 2007, I wanted to focus solely on SQL Server and get out of the IT/help desk type role, so I took a job at Sogeti. SQL engineer was my title, and the rest is kind of history. I've been doing data related SQL Server stuff since 2007.

Eleven years, and now I'm here. It was actually a really good company. I worked for them for about three and a half years, learned a lot. I had some clients in Omaha that really taught me a few things. I got lucky. I had some really good clients, one of which was Farm Credit Services of America. In July 2010, I made the decision to join them full time as a database developer. After I was at Farm Credit Services of America for two years, I took a position as a database architect. Eventually, I made my way to Louisville working for Farm Credit Mid-America. Farm Credit Mid-America and Farm Credit Services of America have a very symbiotic relationship. Ironically, the CIOs of both organizations are brothers.

I got to work with both of them when I was in Omaha, and one of them left to come down here and be the CIO of the Farm Credit Mid-America here. And fast-forward a couple years, I started the conversation, but they were kind enough to recruit me. That's how I ended up in Louisville, Kentucky.

Mala: Describe a few things you wish you knew when you started your career that you know now and you'd recommend to people that are new.

John: That's a good question. Probably one of the things I always try to talk about is getting involved with the SQL community. You're very active in the SQL community, and I wish that I had figured that out a little earlier. Whether that's blogging or getting on Twitter or speaking. I think speaking and doing presentations is something that anybody, especially in the IT career, needs to know how to do. I can fully say that being involved in the community has been a blessing and a strong push in my career, as organizations look for people that are speakers and active in their fields within their IT community. So I wish that I had known.

Being involved in a community was an aspect of my career that Farm Credit Mid-America really liked, and they wanted to get involved in the community. I did have to have those conversations with them. I want to go speak. I want to be involved in a user group if I can. Help at SQLSaturday. I was blessed that Farm Credit Mid-America was very willing to allow me to continue to do that.

Mala: You're a big proponent of agile, so describe how a DBA manager maintains his or her policing/gatekeeping duties with frequent deployments that come with the environment.

John: Doing agile with a database is sometimes like dancing through a minefield.

Agile with application development is much easier because it's much easier to roll out application changes and rollback and handle failures. You copy the old build DLL and replace it if something blows up, so rollbacks are really easy. You have to worry about the data. Most of the time, agile methodologies work okay with the database, and you can facilitate that through various tools like continuous delivery or database version management. We really have to be aware as to what changes will break things and what won't.

Just like application development, we should deploy database changes in small, incremental steps. In a former life, I was a database developer on an agile team, and that's what we did. We made changes to the database and those changes were small in scope. As my developers worked on stories, there was usually a database component that I would work on. After a two-week sprint, when they were done, we released all of those changes all at once. I would manage as the database developer to ensure my changes weren't breaking things.

Mala: Do it in small increments and be part of the development process?

John: Yes, that too. I would also say that having a testing process is very important. At least from a data professional side, as we do database changes, we don't really test things. There are some tools out there that can help facilitate doing unit-type tests against the database, but we don't do it. The unit testing or test-driven development is very common on the application side. It's like a second nature to them. From a DBA perspective, we don't do that.

To help facilitate those small incremental changes, we should also look at implementing some type of testing harness to facilitate that. Redgate has tools to do some of that testing methodology.

Mala: Now, going further into that, what are the tools you recommend for database version control? What have you used that you really like?

John: The two major ones that I've used are SQL Server Data Tools, which is SSDT, and the Redgate source control that plugs into SQL Server management studio.

Both have pros and cons. My personal preference is the Redgate tool for a really selfish reason. Because I'm a DBA, I spend the majority of my time in Management Studio, so I can facilitate the source control of a database right there in my backyard, basically. While Management Studio's based on Visual Studio framework, it's a different sort of beast, and it's got its own sort of nuances in terms of putting the database in the database project and managing it within Visual Studio. It can be done. I've done it. In either case, I do think that databases should be in source control, which seems to be a stigma that most DBAs don't get. I prefer the Redgate source control product for database source control.

Mala: Data science has been kind of a game changer, the way that data roles are going to evolve. How do you think a DBA's role is going to evolve with agile? Do you think there will be more specialized jobs that happen because of more and more people adopting agile?

John: I don't think that role is going to necessarily change. We are still the stewards of the data, and if your role is an architect-level role, you're still going to be involved with how to design and construct the database, as well as being involved with the deployment of those changes. One of the things that I think will change is that as more organizations adopt things like database lifecycle management, and implement tools, database delivery will allow smaller changes at a faster pace. That might help the database role become easier.

As a DBA, if I have the tools in place, and I trust the process and the tools, I actually empower developers to make database changes in a development environment, and they can control the process in which we actually release to the product.

Agile and database lifecycle management should, in theory, make the DBA's life easier, because usually, most deployments go barf on the database deployment, right? That's usually the trickiest part. Changing the database under the hood causes problems. We delete data. We destroy data. Things go bump, and we have problems. If we have a process in place and it's agile, then the DBA's role will actually be easier. In theory, though.

Mala: A lot of DBAs I've talked to don't like being part of stand-up meetings, or the concept of one DBA per team kind of approach. What has been your experience with that? Some people think it's too much overhead for them.

John: I think that really depends on the culture of the organization. I am a fan of having some routine stand-up meetings. I think it really depends also on the actual role. When I was a database developer, we had stand-up meetings. In that role, I actually sat with my application development team, sat right next to the QA people or the BAs or the developers, and we had a stand-up every morning. I think that was beneficial because we knew what pace and who was working on what. From an operational standpoint, I don't think that a daily stand-up is necessary.

I have been in shops, the Farm Credit Services of America in Omaha, when I was a database architect, I was one of four, and the four of us supported I believe ten application development teams. We weren't assigned to any one team, and really the theory was each DBA had two teams to manage. That's kind of how they broke it down. We did have stand-up, and we did our work in an agile-type methodology. Even from an operational standpoint, we had a Kanban board, we did sprints, we estimated in hours how much workload can we take outside of the day-to-day support tasks, what projects could take on in a two-week sprint and get done? We actually had a stand-up once every two weeks, and that cadence was really comfortable from a DBA perspective. We knew who was working on what for the next two weeks from an operational standpoint, and that also facilitated. We weren't just bogged down with meetings every day. We all sat together, so we all knew what was going on anyway. A stand-up scheduled every two weeks was about right.

Mala: As a DBA, what sorts of issues are caused when you interact with other technologies like site admins? And who are you most likely to clash with and why, how do you handle that?

John: That's a great question. I have clashed with every single one of those, in part. I'm a DBA. I've had experience with developers, SAN administrators, network folks, management, QAs, you name it. I've interacted with and clashed at some point in time. What I've experienced is if your DBA has knowledge around those areas, especially with the SAN admins or the network folks, if you go and talk to them, usually I find that they're somewhat surprised that the DBA has some type of knowledge level in their area of expertise.

I've actually had to go to SAN administrators with a problem. Usually, I get some type of pushback, but we're all professional and we're all there to do the same job. We want to make our employer successful because that's what gets us. Understand they're just doing their job, and have a friendly conversation about how to fix a particular problem.

I am probably most likely to clash with developers. Not that I'm picking on developers, but usually they're designing something that doesn't quite work with SQL Server. We have that conversation about how to do it differently, which is usually a paradigm shift for them to some degree.

Mala: What has been your experience as a DBA, and then as a consultant, with cloud adoption?

John: Oh, cloud adoption. My last two employers, because they're financial, cloud adoption from a SQL perspective or data perspective has been very, very slow, and rightfully so. You don't want to put customer data into someplace that they could not physically control for data protection, obviously. As a consultant, working with the cloud was actually one of my goals. I see the cloud becoming very prevalent in the IT world. I wanted to experience that cloud work so I could get some experience with it. A lot of my clients now have some type of cloud presence in some fashion. Whether it's, maybe not SQL Server, but they've got file servers in the cloud. They've got application servers in the cloud. They've got data storage in the cloud somewhere. Cloud adoption is just going to increase over time as Microsoft and other vendors get out there more, and more technical under the hood. Cloud adoption's just going to increase, I think.

Mala: If a new customer comes to you and says they want to come to the cloud, what's the approach that you take?

John: I don't want to lead them down the wrong path since I'm still new to the cloud adoption. I like the Microsoft products. I don't have any experience with AWS or the other providers. I'd ask the client what they are looking to do in the cloud and how can we get them there. Obviously, there are migration tools, so we can migrate data from their on-premise instances of SQL Server to the cloud and get them where they need to go.

Mala: What are some industry trends that you're particularly excited about and watching?

John: Right now, I think the industry trend of data privacy is really interesting. Now what's going on is that whole Facebook debacle, where all that data was disclosed to a third party unknowingly, or however that works. I don't know. I think data privacy and data security is going to be really interesting to watch and see what comes out of that. I foresee some more legislation—like GDPR [*General Data Protection Regulation*] in the EU, the European Union—that the United States adopts some type of the same laws. I think that'll be interesting to watch as well.

Mala: What's the importance you place on documentation for the DBA, and how do you approach it?

John: I think documentation is very important. To be fair, I am not the best at it. I've gotten better over the years, and the way I approach it is I use tools like Evernote or OneNote. We use OneNote here at Denny Cherry & Associates, so almost everything we need to make notes of goes into OneNote, where everybody has access to it. This allows us to collaborate across the board on how to handle clients. Or if things come up and I'm out of communication, my colleagues can pick up those notes and go do what they need to do.

I don't think that DBAs, in general, document stuff very well. I think that's a difficult task for probably any type of IT professional. I just try to make sure that I have a mental note, "Oh, I should write this down," that I then actually go to Evernote or OneNote and make a note, whether it's a screenshot or just a snippet of code. I try to make sure to do that. That's really a habit thing.

I don't think we do a very good job of it, unfortunately.

Mala: Most people find it boring.

John: Yes, it is boring.

Mala: What are your favorite books, blogs, and other means of learning? How do you keep up?

John: IT technology is exploding so fast. I think it's difficult to keep up. I really do. I'm not much of an IT book reader. I am currently trying to study for a certification exam, but I usually learn by practice. Blogs are a huge part of my learning process.

I also tend to share things. If I come across something from a client, we have an internal chat. We use teams. We have an internal chat window. I leverage the knowledge of my colleagues on A. And my colleagues, who are all very, very smart, usually tell me that I'm completely wrong or completely right. One of the two. Talking and leveraging my coworkers is a huge way for me to learn.

Mala: You were recently awarded the MVP. What has been your experience with the program?

John: Since being awarded the MVP award, I do enjoy the program. Obviously, there are things that I can't discuss for non-disclosure reasons, but I do get exposure to the newer technology that might be coming out, and I've got access to some of the really smart people at Microsoft who build the products. I can actually have a direct line to them, and that's helpful.

Mala: Are you liking your experience with it? I was looking more at the experience you got after you got.

John: You get some free stuff for being an MVP. The real experience is getting access to the early release of new technology and finding out how things work. There are some distribution lists that if I run into a problem, I can reach out to other MVPs. Those lists are watched by the various program managers of the respective products, and so I can get answers back relatively quickly from Microsoft. So far, I've enjoyed it, but I've only been an MVP for a few months.

Mala: Are there any tools and techniques that you use on a daily basis as a DBA that you really like?

John: There's probably a handful. The one I use the most is Snagit by TechSmith. I am forever taking snippets of various screenshots and sending them to clients. That tool has been invaluable to me, so I visually show my clients what I'm looking at or whatever I'm talking about. I bet I take fifty screenshots a day. That thing is always running in the background. I can easily take screenshots of various things, Mark it up and send it off. That tool has been fantastic.

I use Trello for personal stuff as well as my professional stuff. That tool, in particular, helps to keep me in line with what I've got going on. I like the visual depiction of moving cards through a swim lane to the done column, and so I can visually see where I'm at with each project or each task.

The other one I've used is Toggl, which is a paid service, but the free version works for me. It's just a time tracker. A mobile app. And some other plugins for various browsers, so I use that to keep track of my billable time for my clients. That tool has been very useful in making sure that I track accurate time for my clients.

Those tools are probably the tools that I use the most.

Mala: Are there any SQL Server tools that you like? Free tools?

John: SQL Search is a fantastic free product. I really like the Redgate prompt. That's not free. I don't even know what that one costs anymore. Notepad++ is always a favorite.

I do like Mladen's SSMS Tools Pack. If you're running something earlier than 2012, that's a great tool to have. That's actually one that I'd consider, even the paid version is inexpensive. I've thought about purchasing the paid version just to have it because he's got some really cool things in there that are very versatile.

Plan Explorer from SentryOne is a fantastic tool.

Mala: Are there any training conferences that you're particularly fond of?

John: Obviously, the PASS Summit is always one of my favorites. I've gone for the last seven years. SQLSaturdays are always good conferences that I enjoy. I'm trying to share and do a little bit more with IT/Dev Connections.

I actually submitted to speak at those, so I'm hoping to get selected. I'm trying to branch out a little bit and do some very different conferences to get some different exposure.

I really like PASS Summit. We all know the power of the SQL family.

Mala: What are your recommended ways of stress management and developing work/life balance?

John: One of the ways I've been doing stress management is when I joined Denny Cherry & Associates, I bought a stand-up desk for my home office, so I can stand up. I have a stress ball, but I also have resistance bands. When I'm on a conference call, usually when I'm muted, I'm not talking, I'm just listening, I will stand up and move around and do exercises, so that helps me with making some type of healthy work/life balance, as well as helping get rid of some of the stress. If I'm stressed out, I can actually use those resistance bands to do a little exercise, and for me, that helps.

Denny Cherry & Associates is very easy to work for. I'm not saying that we don't work hard, but the work/life balance for me working from home is a very nice change of pace from an office. I spent most of my IT career working in an office. This is my first exposure to working from home full time. Really, as long as the work's getting done, having the flexibility to have that work/life balance is a really good thing for me.

Yesterday I spent some time with my son in the afternoon, but this morning, I knew that I had to do some work, so I worked for a few hours. I'll finish up the work later and as long as my clients are happy, they're okay with that.

Mala: Describe your style of interviewing a data professional. What are some examples of questions you ask?

John: I love interviewing data professionals, especially DBAs. What I look for are a couple different things. One, I want to see how deep their technical knowledge is. I will ask relatively complex questions about some of the internals of SQL Server if that job entails SSIS development. I might ask some technical questions about SSIS or SSAS. I want to ask them questions until they can tell me they don't know. I want to hire somebody who has the ability to tell me that he or she does not know, but they know where to go to ask.

I try to decipher if they are aware of the community. Do they know how to ask questions on the forum? Do they read blogs by Denny Cherry, or Brent Ozar, or Paul Randal, or Kimberly Tripp, or Kendra Little? Are they aware of all those great resources out there on how to find answers to the technical questions? Those are the kind of questions that I look for. I also look for culture fit, depending on what the culture of the organization is.

Mala: That's important.

John: I think that's very important. I think that I would almost say that culture and personality are a little bit more important than the technical skills because we can teach the technical skills. I can teach it, they can go to classes, conferences, whatever, but a good culture fit is really vitally important.

Mala: What are your contributions to the community, and why do you recommend people be involved with the community?

John: I think that I contribute a lot to the community. I am a speaker at SQLSaturday. I have submitted to some other conferences. I do blog about SQL Server. I have led two different user groups. I led the Omaha SQL Server user group for seven years, and now I help lead the one here in Louisville.

I think the community is a huge benefit for individuals to promote their career. I don't think that I would be where I am now in terms of being here in Louisville, as a speaker and MVP, if I had not got involved in the community many years ago. I think that being involved in the community is a huge boost to a career. You should get involved with the community, whether it's SQL Server, or Java, or whatever, people should get involved with their local technical community. If there isn't one, start one, because I'm sure there are other individuals like you that want to get involved, but there is no community.

Mala: Can you narrate a funny, interesting story to share before we wind up?

John: Yeah, I'm trying to think. I've got one from my Lotus Notes days that I tell at times. Lotus Notes is an email platform, and back then, I was the Lotus Notes administrator, so I controlled all the incoming mail, and for whatever reason, the organization I worked for, all of our spam mail actually got delivered to us, and we handled it. Keep in mind this is fifteen years ago. A friend of mine, we had this little game where when we walked by each other in the hallway, we would just kid each other. Nothing graphic. Nothing horrible. One day my friend walked by and he called me a name. I was totally cool. I wasn't mad or anything. However, I went back to my desk, and I changed the email spam filter to deliver all the spam to him.

Within about fifteen minutes, he had about eight thousand messages in his mailbox, and they were all spam. They were all junk, but if you've ever used Lotus Notes, what's funny was, if you selected a bunch of messages and then hit Delete, Lotus Notes would process that delete, and behind the scenes, it would continue to accept new messages. He was in the process of deleting all these messages, which took him a few minutes, and by the time he deleted those messages, another three thousand had showed up.

The other part of this story is that he actually came over to my desk and had asked me if anything was going on with the email and the spam. And with a straight face, I told him nope, everything looks fine. I let him go back to his desk, and another five thousand messages had shown up. At some point in time, I turned it off and let him clean up his mailbox, but that's always been funny. Don't mess with the email administrator or the DBA. He'll direct all the spam to your mailbox.

Key Takeaways

- Doing agile with a database is like dancing through a minefield. The unit testing or test-driven development is very common within the application side. It's like a second nature to them. From a DBA perspective, we don't do that.

- As a DBA, if I have the tools in place, and I trust the process and the tools, I actually empower developers to make database changes in a development environment, and they can control the process in which we actually release to the product.

- As more organizations adopt things like database life-cycle management and implement tools for database delivery that allow smaller changes at a faster pace—that helps the database role become easier.

Favorite tools: Redgate source control product, SQL Server Data Tools, Snagit, SQL Search, Notepad++, SSMS Tools Pack, SentryOne Plan Explorer

Recommended conferences: PASS Summit, SQLSaturday, IT/Dev Connections

Kathi Kellenberger

Editor, Simple Talk, Redgate Software

Kathi Kellenberger, *is a Data Platform MVP. She is also the editor of Simple Talk, Redgate's online technical journal. She began working in technology in 1997. Since becoming a SQL Server DBA in 2002, she has focused on data. In addition to working at Redgate, she owns Kellenberger Consulting, LLC, writes books for Apress, and creates courses for Pluralsight. Her proudest career accomplishment is writing Beginning T-SQL (Apress, 2014), which has helped people around the world get started with SQL Server.*

Kathi has a bachelor's degree from St. Louis College of Pharmacy and a master's degree in computing and information systems from Southern Illinois University. She currently volunteers as an instructor in the CoderGirl program at LaunchCode, which provides free training and matches students with corporations in paid apprenticeships. She is also co-leader of the PASS Women in Technology virtual group, and she is on the board of the St. Louis SQL Server User Group. She has volunteered at PASS in several roles since 2005, including serving as a program committee manager, which led to her winning the PASSion Award in 2008.

© Malathi Mahadevan 2018
M. Mahadevan, *Data Professionals at Work*,
https://doi.org/10.1007/978-1-4842-3967-4_18

Kathi loves singing, especially karaoke. She organized the first "SQL karaoke night" in 2006. Her favorite athletic event is stair climbing. She has participated in several stairs climbs in St. Louis and climbed the Space Needle (timed at 13 minutes 11 seconds) in 2017. Kathi loves spending time with her friends and family, especially her five grandchildren. She can be reached on twitter at @auntkathi.

Mala Mahadevan: Describe your journey to the data profession.

Kathi Kellenberger: I took a very strange route to the profession. I started out as a pharmacist and eventually became a developer. It was Visual Basic, mostly. In 1997, I left pharmacy and became a full-time developer. I was planning on continuing that route, and what ended up happening is that I was a developer on a project at this big law firm. I really liked the people there, and it was an easy commute. When a job opened up for a DBA, I applied and got the job. That was in 2002, and data has been my focus ever since. So I think it was a great move—just something I hadn't planned on happening.

Mala: So since then you've been working as a DBA, and then a consultant, and then now an editor, right?

Kathi: Yes. I was a consultant at Pragmatic Works. I was on some DBA projects but mostly business intelligence. And then I left there in 2014 and ran my little Kellenberger Consulting business, which I still have, but it's nothing like it was the last few years. And that work was a lot of SQL health checks and query tuning—things like that for small companies. I wasn't really a database administrator anywhere, but sometimes I would just go in and see what's wrong with the SQL Servers, and then make some recommendations. And I may or may not be the person who carries those out.

Mala: Describe a few things you wish you knew when you started your career that you know now.

Kathi: As a teenager, I had my heart set on one career that ended up not being a great fit for me. I just wish that I had realized earlier what I would be good at.

Mala: If you had to do it again, how would you go about figuring out what you're good at? I've heard that line from many people, and I'm just wondering what we could tell people who are younger. How does one figure out what you're good at early on so that you don't have to learn by error all the time?

Kathi: I went to a college that was specific for pharmacy right out of high school. And of course, this was a long time ago, before the Internet. There are more resources now, but I feel I should have gone to a regular university so I could have easily changed majors. I just felt like I was kind of stuck on the path I was on. If I had gone to a university and taken a variety of classes the first couple of years… If I had taken any kind of programming class, I would have known right then.

Mala: So education matters. Like if you learn different things, you're probably going to find out.

Kathi: A lot of tech people are very successful and don't go to college. I just think it's a matter of being exposed to different things. Once I was working as a student in a retail pharmacy, I did not like it at that point and realized I had made a mistake. I had never seen computer code. As soon as I saw computer code, a little program on the TRS-80 when I was almost done with pharmacy school, at that point, I was hooked. I just had never had that opportunity until then. The high school I attended didn't have any computers.

I think when you finally find that thing you are meant to do, you really know it.

Mala: So you made some major career decisions, moving from different areas of work in the data profession—first from a DBA to a consultant, and now you're with Redgate as an editor. So what's your advice for people who are pondering similar things? Like moving from being a long-term DBA to something else?

Kathi: I've actually always had a regular job, but on the side, I'm doing some other thing that interests me.

For example, when I was a pharmacist, I started doing web development and database development on the side. I was able to get a career as a developer after a couple of years. Switching from being a database administrator to a consultant was not a difficult move. With consulting, the challenging part is that you are constantly dependent on how long the projects are. You're going into a new place all the time—where you don't really know the environment and you've got to figure it out pretty quickly.

But as far as being a editor, it's been really great because I started writing articles and books around 2004, and I just continued to be involved with writing and editing, and so now that's what I do. That's my job.

So it's pretty amazing. But it seems like if you can work on whatever it is that you're interested in, and get some experience, then it might be possible to make the switch.

Mala: What's a typical day in your life as an editor at Redgate?

Kathi: Oh, it's interesting because I work from my home office. I don't have to do a lot to get ready for work, and I don't have to drive. I can just walk into my office, but it is important that I start working kind of early because a lot of my coworkers are in Cambridge, England. All the meetings are before noon.

A typical day for me is checking to see all the articles in progress—which ones need to be worked on. Is there something that I'm close to having reviewed that I can send back to the author? Do I have something from an author that I need to take a look at? I'm doing a lot of editing all day. I'm also writing editorials and preparing for speaking engagements. So that's my typical day.

Redgate is really flexible. If for some reason I decide I am going to take off an afternoon and play with the grandkids or whatever, I can do that, and then just pick it up later in the evening.

Mala: You read a lot of articles written by a lot of very talented people. So obviously you're well informed on things that are going on. What are some technology trends that you're really excited about?

Kathi: Some of the articles that I've had to review talk about the cool things that you can do in Azure with very little coding. Workflows. There was an article that I tech-reviewed recently on creating a bot that automatically answers typical questions. I see more and more people moving things to Azure or AWS [Amazon Web Services]. It's just amazing functionality. Basically, you have these pieces that you can customize to do what you need to do—and not worry about servers or infrastructure. It's really, really cool.

Mala: When you were a DBA or a consultant, what were some of your favorite tools and techniques?

Kathi: I was a database administrator from 2002 to 2010. I didn't have a lot of extra tools back then. I had Idera to monitor some of the databases, some of the servers. My main tool was scripting. I just scripted things like crazy.

Mala: What was your favorite scripting language?

Kathi: T-SQL is my favorite. I think back then it was all T-SQL and VBScript. Since then, I have done more and more with PowerShell.

Mala: What sort of struggles have you faced with management or the business side of things in any position you've had? What was your approach to handling that?

Kathi: Ah, this would probably get me in trouble. When I was a database administrator, I kept getting more servers to care for. It always was increasing.

I really wanted to have a DBA team. I had one person—super talented in so many things. I had about a third of her time to help me. And then she would kind of take over if I was on vacation. But I really never got through to them that I needed help—that we could be so much better if I had a two- or three-person team. And now they do have several people on a team. So that was a major problem I had. I was on call 24/7.

Mala: Oh, my. That's hard.

Kathi: It was hard. I was really the first full-time DBA they had. I had things pretty well ironed out. Over time, there were a lot fewer calls and a lot fewer problems that were happening at night, but still, I had to make sure I was always available.

I loved the job, though, and I loved the company, but, at least having one full-time person to work with would have been great.

Mala: So if you had to do that job again would you take a different approach to telling them they need more people?

Kathi: I don't know. Now they do have a DBA team. Yeah, if I could figure out what that approach was, I would have done it. I just never figured it out.

Mala: Did you ever run into issues interacting with other technologists? Developers, SAN admins, or network folks.

Kathi: For the most part, I got along really well with them if they understood what I was trying to accomplish. I had one crazy thing that happened. This was the first time I was setting up clustering, around 2003. I had to go to this network guy, and I said, "I need IP addresses. I need you to reserve them." You need a bunch of IPs whenever you're clustering, but he didn't believe that I needed the IPs, and he said I couldn't have them. I had to take the instructions on clustering over to him to show him that I needed these IPs. And then after that, I really didn't have a problem with it.

And the other thing that was kind of aggravating is that different IT team members would just say, "I need your SA [system admin] password for this server." One time it was because of one of those backup companies that pull backups right out of the server. It seems like it was Backup Exec. It could pull the backup straight from SQL Server right to tape. And they wanted to set that up, but I didn't want that since it would make me dependent on another team for restores. The reason they thought they needed to do it is because the Backup Exec sales guy came in and told them that, unless you're using their tool or shutting your SQL servers off, your databases are not getting backed up.

And so they were convinced that the databases were not getting backed up. And I said, "Oh, yeah, they are. I'm doing native backups, and then you guys are copying the backup files to tape." They were adamant that they needed SA passwords for all the servers so they could set up this Backup Exec job. And they hadn't told me about it before then or invited me to the meetings about it or anything. But, I won that battle. We just kept doing things the way we were doing, and it was fine.

I ended up having one server that I called my recovery server. I just automatically copied over all backup files so my backups were in two places, and I could do test restores. Basically, they just backed up the repository on that server, and it really simplified things.

Mala: What's the role of documentation in your current job? Do you have a need to document things? And do you recommend that people do that as a habit?

Kathi: As the editor, I don't have a lot of things to document. As a database administrator, I documented like crazy. In a law firm, you have document management software. Whenever you try to save a document, a profile screen comes up, so you have to assign categories. It's really great. And you have automatic versions of documents. It was a fantastic way to create documentation. No matter what it was, I had a document out there explaining it. The real challenge was keeping things up-to-date.

Mala: What are your favorite books, blogs, or other means of learning? How do you keep up?

Kathi: My favorite books are written by Kalen Delaney or Itzik Ben-Gan. Early in my DBA career, I happened upon SQLServerCentral. I had that page open like every day and learned a ton there. I often find myself on the SQLskills site or on Kendra Little [SQL Workbooks]. I really don't have any particular one that I'm going out and checking every day. It's a matter of searching, and then I end up landing on some of the same resources.

Mala: Are there particular conferences or other ways of learning that you really like?

Kathi: Over the years, I do less learning at conferences. I'm speaking and networking. It's kind of rare for me to actually attend any sessions. At SQLBits, I went to two sessions by Bob Ward. At DEVintersections in Orlando, I went to a Paul Randal session. And there are all these fantastic sessions, but I end up spending more time meeting with people, talking with people. You know, making connections.

And then, of course, at PASS Summit, there are recordings, so if something comes up, I'll go back and look at the recordings.

Mala: What are your recommended ways of stress management and work/life balance? How do you handle it?

Kathi: I try to really plan out what I'm going to do each day. I try to have one day every weekend where I just do family things. I also do a lot of work for a volunteer organization. I might say, "Okay, on Saturday, I'm going to work all day on Women in Technology [WIT] stuff and LaunchCode stuff, and on Sunday, I'm taking my grandkids to the zoo." I also really, really enjoy just soaking in a hot bathtub. I look at the clock and say, "Oh, my gosh. It's nine o'clock. I'm stopping. I'm just going to soak in the tub and read fiction." That's pretty much what I do.

And I also like to get out and exercise—riding bicycles and walking.

Mala: You've been a longtime proponent of the WIT cause. I hear from a lot of women about family time and work/life balance, especially as a DBA and data professional. Do you have anything in particular you would want to say to women in that regard?

Kathi: I know it's very difficult. I was lucky that when I switched to this type of career, my kids were a little older. I became a full-time developer in 1997, and my daughter graduated from high school in 1998. So she was already in high school and my son was just about to go to high school. So it was easier for me.

I think you've got to have a team at home. My husband and I are a team. And there's housework. You know, whatever it is that needs to be done, one of us has to do it. He's great with cars, and I don't even like to drive them, so he takes care of them. We are a really good team. From day one he was the one getting up with the babies whenever they were crying at night. If you have a partner that is willing to share responsibilities with the kids, the household, it's going to make your life a lot easier.

Mala: So you've been a long-term MVP, too. What are your positive experiences with the MVP program?

Kathi: I have been an MVP for a long time. I was first awarded in 2008. It's really an honor. It's a fantastic program.

A couple of the benefits that I really love are the MVP Summit, and the MSDN subscription. I think the program is very cool because if you need to talk to somebody on one of the product groups, being an MVP, you can do that. When I was working on the very first Integration Services book back in 2005 with Brian Knight, he gave us names of people at Microsoft that could help us if we ran into issues. And I never got even one of my emails answered.

More recently, I wrote a book on Reporting Services, and I created a Pluralsight course on Mobile Reports. I was able to ask questions easily of the team that worked on these products and got answers back quickly. It was really great. So I think that's another benefit. You can have access to these people at Microsoft who would otherwise maybe not answer you.

Mala: So as an extension of that same question, why do you recommend people be involved with the community?

Kathi: It really can change your life. I started getting involved with the community in 2004.

When I really started getting involved with the community, I decided to volunteer. In 2008, I became an MVP. That was only four years later. But it leads to opportunities. It gives you credibility.

If you're writing and speaking, that's really going to improve your opportunities. If it's organizing a SQLSaturday or volunteering locally, those are wonderful things, but I'm not sure if they actually lead to as many opportunities. You make a lot of connections. Staying local doesn't really get your name out to the world per se like writing and speaking do, but they are still worthwhile activities.

The number of relationships you develop is just amazing. It's like every city in the US, I have friends. It is a wonderful experience getting to know people and having these friends all over the place.

Mala: To wrap up, is there a funny story you want to relate? It does not have to be work-related.

Kathi: My life is so centered on grandchildren. I have five grandchildren. One of them is a granddaughter, Gwen, and she is six. Last year, during one of our many trips to the zoo, she decided that the Galapagos tortoises—there were actually two different species, but these giant tortoises were her favorite animal.

So the next time I took her, she wanted just to sit there and watch them. And then she turned around after twenty minutes and said, "Grandma, can I play with your phone?" She just wanted to sit there and watch them, but she was still kind of bored. It seemed funny at the time.

In Hawaii recently, my husband and I renewed our wedding vows. Immediately after our little ceremony, she asked me if we were having kids!

Mala: Kids are always funny.

Key Takeaways

- When you finally find that thing you really want to do, you know it. Being exposed to different things helps.

- If you can do some of the work you're interested in on the side and get some experience at it, then it might be possible to make the switch to doing it full time.

- Women in particular have to have a team at home. If you have a partner that is willing to share responsibilities, it's going to make your life a lot easier.

Blogs to follow: SQLskills.com, SQLServerCentral.com, Kendra Little's SQL Workbooks blog

Argenis Fernandez

Principal Data Storage Architect, Pure Storage

Argenis Fernandez *is a principal data management architect for Pure Storage, based in the greater Seattle, Washington, area. He has worked with SQL Server since 1998 and enjoys working with large SQL Server farms, busy databases, performance troubleshooting, high availability, disaster recovery, best practices, and PowerShell scripting.*

Prior to Pure Storage, Argenis worked as a lead database operations engineer at SurveyMonkey and as a senior consultant on SQL Server Core for Microsoft Consulting Services. In 2013, he founded the Security Virtual Chapter for the Professional Association for SQL Server (PASS) (http://security.sqlpass.org). He also handled the SQLSaturday portfolio for PASS as part of its board of directors in 2016.

He has been awarded as a VMware vExpert for several years in a row.

M. Mahadevan, *Data Professionals at Work*,
https://doi.org/10.1007/978-1-4842-3967-4_19

Argenis is a SQL community enthusiast and speaks frequently at major SQL Server conferences, including the PASS Summit, PASS SQL Rally, IT/Dev Connections, SQLBits, and TechEd. He is also a Microsoft Certified Master on SQL Server, an avid Twitter user (you can follow him at @DBArgenis), and occasional blogger on SQL Server topics at SQLBlog.com. He also blogs on more storage related topics at https:// blogs.purestorage.com/blogs/argenis.

Mala Mahadevan: Describe your journey into the data profession.

Argenis Fernandez: It's twenty years in the making really. My first job out of college, I was working for an ISP down in Venezuela. It was basically the AOL of Venezuela. It was called CANTV. It was actually the local phone company. They had a monopoly in Venezuela. It was actually owned by Verizon back then. It was interesting, straight out of the college without a lot of experience, and I only had the experience that I gained at the computer lab at college, nothing more than that.

I started working for them as a Unix admin. I was basically doing sysadmin stuff for them. I was very much focused on infrastructure, I wasn't focused so much on data. Then, during that time, working for that ISP is when I actually got thrown into the Microsoft world and the Windows world, because the company made the decision to migrate their systems to something called Microsoft Commercial Internet System. That's something that wasn't commercially available. Well, it was commercially available, but only to ISPs and ASPs, like application service providers.

Way back when. We're talking 1998. That's how I got thrown into the deal with Windows as a serious operating system to handle server stuff. That was my first interaction with SQL Server. MCIS, that was the name of the system, was based on SQL Server as a back-end. It used Site Server for an LDAP server, the precursor to Active Directory. Microsoft was starting to manage all these technologies and developing them.

Eventually with Windows 2000, this became Active Directory. That was the first time that I ever worked with data seriously. It was managing the data for more than two hundred and fifty thousand people. Imagine, that was a big responsibility. I moved to the United States after that job. I started working for a couple of startups. I was always the database guy because there was basically no one else who knew anything about databases.

After a few years of doing a little bit of everything, I focused seriously on SQL Server. I dedicated myself exclusively to SQL Server in 2004. That's all I've ever done since. I went through several DBA jobs. I ended up supervising a team of IT professionals, but I was basically the database guy. That was back in 2006, working for a newspaper down in South Florida. After that, I did a couple of more jobs in South Florida, then I ended up here in Redmond working for Microsoft as a DBA.

That first gig as a DBA for Microsoft was really interesting because it was a massive amount of data. We're talking literally terabytes and terabytes and terabytes. Actually, in 2009, when I was considering moving to Seattle, I really put my sights on the SQL Server community. I really put a lot of focus into expanding my reach into the community, speaking at SQLSaturday, not just attending it. Even if I couldn't speak at it, I would really put a lot of effort into making sure that I was there.

Being there is a big part of it. A lot of people just submit and keep away from the community, and that's okay. Most of the people that are really successful out there in the community are actually there all the time. You see them around all the time. To be present is a big part of it. That is a big part of it. After moved up here to Redmond and a few years, I got my MCM. I put a lot of work into getting the MCM certification. I went back to work for Microsoft for MCS. Then I left. I worked for SurveyMonkey, where I got my MVP and eventually ended up here at Pure Storage.

I've always kept an eye on infrastructure. I've seen such a big, radical change in the way that people approach infrastructure. In the past, everything was very static. You will buy some servers and you will buy some network and some storage. Then you would never consider even rebuilding a server.

You remember how painful that was back in the day? Rebuilding a server was a whole ordeal that will take you a whole day. Nowadays, when you want to build a server, if you think about how infrastructures work, it's pretty much stateless. You just deploy a server in seconds from a profile that's saved on some storage. You just deploy it, and in a matter of minutes, you're up and running your workload on that server. Whereas, years ago, everything was very static.

Just to have some compute available, you had to go through this big exercise of buying a big server, spending hundreds of thousands of dollars on it. Then installing the operating system will be somebody putting CDs into the DVDs. You will boot up a CD to get your operating system installed. Everything has changed so radically. Everything is push of a button these days.

People should really think about that. A lot of the manual tasks that you do today can pretty much be considered obsolete. The only way that you can guarantee that you're going to have a stable job and a good progression of your career are by making sure that you always had made as much as possible. Always try to deliver value in terms of what you do, as opposed to just being seen as a cost center. I think that's just good for people to focus on.

Mala: You've talked about why you're using some commonly used scripts to measure IOPS. And proof scalability is really not the best thing a DBA should do. Can you explain why and what should a DBA really do to evaluate if he or she is getting the best storage they need?

Argenis: What a lot of people focus on is just running something, getting some numbers and making sure that they run those numbers by the storage person, whoever that happens to be in that company. That's really it. Storage validation is a very, very, very complex topic. Working with storage here at Pure, I've actually come to appreciate how difficult and how hairy that stuff can get. Let me give you an idea.

Most storage solutions perform some sort of management of the data, so they perform some handling of the data, like date zero suppression, pattern removal, deduplication, compression, or even encryption of the data. A lot of storage solutions are providing value that way.

They're trying to minimize the amount of footprint that you consume by that data. The explosion of data is a big problem out there. We have some customers that literally have dozens of petabytes of data on their data centers. When you think about how much storage that actually consumes, it's a significant amount. It's a significant investment. It's a big part of the bills. Making sure that they're very efficient in terms of their storage footprint is significant. Anyway, when you test a storage solution that does any of these things, you can't just rely on something that generates pattern data sets. In the end, those patterns are going to handled and basically discarded. They're going to absorb only metadata, and they're not going to really test the underlying storage subsystem and test all the nuances of that subsystem, so that you can actually get an idea of how your workload behaves when that storage subsystem is used.

If you think about how you test a regular hard drive or a regular SSD, it would be okay for you to generate a pattern dataset, because a regular hard drive or a regular SSD wouldn't do any manipulation of data like that. When you're talking about more enterprise or pretty big storage subsystems, more than likely, they're going to be performing some of these things. Sending a pattern data set down to them is basically going to be mainly a futile exercise.

It's only going to really show you how good your connectivity is between your actual compute node or your host, and your actual storage device. If you use things like SQLIO, it's deprecated. Use disk speed, which came to be the replacement for SQLIO. You're basically going to be using a generating pattern dataset by default. The other way of testing generated datasets with disk speed, for example, will be the complete opposite of generating a pattern dataset, which is a high entropy dataset that looks like a compressed or encrypted data set to the actual storage device.

You see how things are basically polarized in that sense. You're not actually sending a dataset that looks like your actual dataset. That's a big disadvantage. You're not actually seeing how your storage device is going to behave when you send it, the actual data set that you're expecting to work with. That is why, when people tell me they're going to perform some storage evaluation and testing, I tell them to just replay the database workload on top of a restored version of the database.

There's nothing better than that, because you're testing with your actual dataset, number one, that shows all the characteristics of the data itself, regarding of little or more patterns, or little or more zeros, or little or more ability to deduplicate that workload, little or more ability to compress that workload. All of that is what we call the reducibility of the dataset itself.

Multiple workloads are going to show multiple degrees of reducibility. They are going to show different behaviors on different storage devices. It is important to focus on the replay portion of testing, rather than just trying to generate some synthetic workload and that's it. Now, a lot of storage companies do the work for you, and can tell you, "Well, we have performed some testing and this is how it goes." They're going to see when you're wrong, different datasets with different degrees of reducibility, etc.

The reality is that you can ask those storage vendors for this stuff. You can tell them, "Hey, can you show me some of the numbers that you've seen for the performance of this database? Let's assume that my database reads, writes, and reduces two to one. Can you tell me how your device will react to that?" They should be able to reply to that. If they are trying to hide those numbers from you, that should be a red flag. The vendor's not really being upfront about the capabilities of their storage device.

What I tell people is, always focus on the actual response from the application, which is your database. It doesn't really make sense for you to focus too much on artificial numbers that may be inflated by the fact that you're looking at a storage device that performs some sort of data management or manages the data itself. If you don't know how to do any of this, you don't feel comfortable replanning workloads and things like that.

There's only one synthetic tool does the work that I think really complies with most of the requirements for a tool that allows you to perform proper stressing validation. That's called Vdbench. It's an Oracle tool. I'll make it perfectly clear that I'm not an Oracle fan, but this is an Oracle tool, and it's free of charge. Vdbench actually works very well for specifying the reducibility of datasets. It also allows you to define multiple IO block sizes, which are important when testing the storage devices. It also allows you to do things like prime the storage subsystem—fill it up before you actually start the proper testing.

To see how it works, you have to do things like garbage collection—when you trigger a garbage collection on an SSD. That's something that people don't think about. They think about an SSD basically works like a disk. It does and it doesn't—it's actually meant to emulate a disk, but you are saving things in a non-magnetic medium. That actually saves multiple bytes at the time and not a single byte at the time, which is a big difference with disk. On disks, if you want to save a single sector, you will save a single sector and that's it. You can modify things by a single sector, and you will be good with that.

On an SSD, you actually modify entire pages. They are 8K in size in most architectures. You actually erase the page and reprogram it every time. When you do that, you are actually wearing down the SSD. If you do a lot of overwrites on SSDs, you're going to end up wearing down that NAND. The actual NAND, the transistors that allow you to save the data. You're going to have to start using some of the other provisioned space that comes from the SSDs.

Some SSDs actually have as much as twenty percent over provisioning. That space that's available to you, in case you wear down the original space. It happens. It happens very frequently. This is how SSDs go bad. If you have a very write intensive workload, your SSDs will wear out, unless you have some software layer on top that ensures that the SSDs are treated differently. Testing different mediums entail that you actually understand what the medium is like. A lot of DBAs don't spend any time thinking about this stuff because they don't have time to really learn from this stuff. They just get a storage device, and then they have to start testing it. It's a little bit of an art. My recommendation for DBAs out there is don't focus on any of that stuff. Run your workload with a realistic dataset or workload, and see what kinds of behavior you'll get from it.

If you can actually stress the workload, meaning running the same workload on top the same dataset, and in a stress fashion, meaning running faster without much thinking time in between statements or generating more threads against the actual database, then you're really cooking with grease. Because at that point, you're really seeing what that storage device is going to do when you actually turn up the heat, and when your customers really go nuts, or when somebody writes a really bad query.

Mala: What are some industry trends in storage with regards to databases that you're excited about?

Argenis: I think some of the most interesting things that I'm seeing with storage or related to avoiding IO, which is something that's very weird. Don't you want a storage device to help you do more IO? Some things that I find super interesting about all these new technologies and arrays, is that you can actually avoid IO for a lot of things. For example, fifteen years ago, if somebody told you to use a snapshot for SQL Server databases, you would be very wary of that.

They wouldn't rely on any of the data services provided by those storage devices. We've come a long way from that. A lot of storage vendors are now providing actual useful snapshots and actual useful clones that don't actually incur into performance issues right off the bat. Years ago, you would experience a severe degradation of performance from even attempting to clone a volume. Because first of all, you'll be running on top of disks so that everything will be super slow in the back end.

The architecture would incur you into hotspots and contention spots that are really negative to your workload. These days, most vendors have great snapshotting solutions and cloning solutions that can have a myriad of different use cases. Let me give you an example. We have customers that take complete snapshots and clones of their OLTP [online transaction processing] subsystem or OLTP system that's customer-facing. That's the one where they're doing all the transactions. They're sticking a copy of the OLTP database into a data warehouse, as opposed to doing an ETL.

Instead of doing ETL across instances, they do an intra instance ETL. They take that snapshot of the OLTP subsystem and they place them inside the data warehouse instance. They just run everything internally. They're shaving down hours of processing that way. It's pretty interesting the things that you can do these days by avoiding a size of data operation, by avoiding IO. I find that super, super, super interesting. Another trend that I see in storage these days, is massive throughput devices. Whereas in the past, you'd be limited to under a gig a second. Even today, the majority of sales that you see deployed out there on storage devices are hard to spec to deliver anything over eight or nine hundred megabytes a second. The reality is that dozens of gigabytes per second is going to become the norm very soon.

That actually enables a whole bunch of interesting things. Somebody on Twitter mentioned the possibility of very high bandwidth rates will be able to perform a schema on demand. Think about that for a second. You have a non-structured set of data, like some logs that you generated from some device. Instead of actually performing lengthy loading into the database while parsing that data and given that data structure so you could query it, using a structured query language like SQL—that's actually exactly what it means, structural query language—you would actually be creating a schema for that data on the fly.

Because at very high speeds, the latency of performing those operations is so low that you would actually be able to afford it. In the future, I think we're actually going to see stuff. I don't think we're too far from that. You just run something and you are getting a view the looks like a database on top of structured data. But the actual storage devices that feed the data so quickly. Plus the speed of the CPUs, and potentially GPUs, help you structure that data from something that has no structure whatsoever. That is very cool.

Then you can apply some interesting techniques on top of that. Machine learning can help you determine whether the processing data is going well or not, like actually smart ETL. You know who did ETL in the past? ETL would tell you, "I loaded all these rows successfully and then here is all the crap that

I couldn't load." There would be a bucket of that. You had to manually go over that data and figure out why you couldn't load it.

You will look at that data and do something with it. You will change the ETL/SSIS process, whatever. The reality is that now we're approaching an era where machines can look at that data and figure out why the data couldn't be loaded. We can end up with smart ETL, smart schema on demand, smart things on the back end. That's actually very, very exciting. People shouldn't be afraid of these things. I see a lot of people shying away from automation and shying away from computers. Hey, they're computers. They have more time than we do. Let's let them do what they're really good at.

Mala: As a storage expert, what has been your experience with cloud adoption? Particularly for SQL Server.

Argenis: I'm seeing a lot of cloud adoption for test and dev environments. I have a very biased view of the market because I work for a storage vendor. I have a little bubble that I work with or realm that I work within. It's very, very, very limited cloud adoption for most of the enterprises that I work with.

I do see, on the other hand, many customers that we don't work with that are just forgoing data center management altogether, and they're just putting everything on the cloud. We don't talk to them, because they don't want to talk to us. I'm a storage vendor, so they have no interest in talking to us whatsoever.

Mala: Do you recognize that approach personally? Like the wholesale movement to the cloud?

Argenis: Not necessarily lift and shift. I'm not a big fan of that. Taking your entire infrastructure, creating VMs off of that and calling that your cloud adoption—that's quite possibly a complete waste of time.

What I call "private cloud" is basically just infrastructure that's on on-prem. You have actually automated it to behave like a cloud. What I call lift and shift will be taking those VMs that you have on on-prem and just moving them to the cloud.

You can run VMs on the cloud—no problem. You can get good performance out of them. You can get them to work. You can get your entire infrastructure reflected into VMs on the cloud. There's no problem with that. The cloud wasn't exactly meant to be infrastructure as a service. Now if there's going be some data store in the back and then all sorts of front end on how you access that data store, which can serve on multiple protocols or can sell multiple database management systems. Imagine that you actually have something that you're accessing via both the Oracle protocol, whatever that's called, and SQL Server's TDS. Imagine that, and applications are actually looking at the exact same data yet something is speaking some language and the other one is speaking something completely different. Yet the data's exactly the same. You can have pretty smart infrastructures. But you can pretty much do that only in the cloud these days, where you have a significant, almost unlimited amount of compute, and not necessarily have all the limitations that you have on-prem.

So, if you really want to take advantage of the cloud you would do that with platform as a service. Not necessarily putting VMs on the cloud. My opinion will probably change next week. But this is what I've seen, customers who have had really good success are the ones that are actually leveraging platform as a service and not infrastructure as a service in the cloud.

I think the cloud is eventually going to end up eating most of the workloads out there, with the exception of very large enterprises, because there's always going to be the sensitive data sets. But I think most people will eventually go to cloud. I mean, most of us are actually going to have to manage some cloud technology in some shape or form. So, we all have to be flexible about that.

And by the way, I think storage right now in the cloud sucks in general. It sucks in Azure, AWS, and in GCP. I think storage in the cloud is going to get really better. Way better than what it is, so we're going to see some big improvements in that area. And that's actually going to help drive an option for everything. If you have slow storage, it doesn't matter how good your computer is. Your performance is going to be bad and your applications are going to suffer. Your users are going to hate it. That's just how it goes.

So, if you make storage easy in the cloud and you make it better, much more performant. I hate that word, but it's pretty much the only one that I can find for describing that very thing. It's just really going to make it a lot easier, so cloud options are going to be driven up significantly. And I think cost is going to come down as well. That's just a fact of Moore's law and also the market itself. I think the cloud is basically just getting started.

Mala: What are some of the ways you recommend to improve the habits and procurement patterns of upper management and other decision makers with regards to storage?

Argenis: Look, that's a hard question. I actually thought about this for a while. Procurement patterns are—typically what happens is that they have established relationships with storage vendors and they get used to working with somebody and they just stick to that one vendor forever.

That actually limits a lot of things that your company could be doing, or how well your enterprise could be doing in general. People become really comfortable with Dell EMC and then they only buy Dell EMC products. Or they become comfortable with HPE and only look at HPE products. Or they only look at IBM and only buy IBM products.

Look, I don't think any company's going to be one hundred percent as successful as they could be by focusing on a portfolio from a single company and this is me speaking as a storage vendor. I want my customers to buy only

Pure Storage, but we don't have a product for everything. They should be looking at multiple companies, and they should be opening their eyes at what's out there. If we look at up-and-coming data recovery companies like Rubrik and Cohesity, those are actually pretty good challengers to the established data-recovery companies.

This is actually healthy stuff. We want competition out on the market, and we want to see many companies vying for customers' money, because if you only have one company that delivers the thing that your company needs, that one thing has no motivation to get better.

That's just capitalism 101. They're not going to throw money at developing a product that no competition whatsoever. So, I think in my opinion companies should be doing a much better job of looking at what's out there and not get too comfortable with buying the same thing over and over.

Mala: So, most IT shops operate on the premise that you're a DBA. You send an email to the SAN guy with the request that I need x gigabytes of storage, and then a few minutes later, you get a response. I've done it a lot of times. That's how we operate. We function in a bubble. What can be done to improve this and help with better capacity planning? How can a DBA be more involved on that side of things?

Argenis: I think one of the biggest problems is that DBAs never actually sit down with the business side and talk about what the expectations for growth are for the business and what the new requirements may be. Or even, whether they're going to have a need for deleting data. GDPR [*general data protection regulation*] is a thing now. If somebody asks you to delete my data, you better be able to delete their data. And you better be able to prove that you deleted that data. So that's actually something that people don't talk about.

GDPR has actually been around for years. It's just that this is the year that's it's actually going to be implemented. So, it's going to be enforced. This is something that people don't talk about, years ago the people who are database architects and DBAs should have sat down with the business and said how compliance is going to affect storage. And do some pretty educated guesses as to where that storage spent is going to be. They should've at least been able to sat down and come up with a basic spreadsheet that has some information on some estimates on what their storage expense is going to be.

Based on whatever it is—business growth, new requirements, acquisitions....If you work for Berkshire Hathaway, you know that you're going to be acquiring companies left and right. Does it make sense for you to consider potential acquisitions during the year as part of your spend? As part of your capacity growth? Absolutely. You should be thinking about all of these things. You know this is part of being a really good DBA or architect. You need to actually talk to the business.

Too often, DBAs are sitting in their chairs, just waiting for requirements to come in. That's actually the wrong approach. The right approach is to talk to everyone. Go talk to your bosses. Go talk to your business, and understand what's actually coming. Sometimes you actually have to force them to speak because they're too busy to talk to you. We're not talking about futuristic stuff. We're talking about planning and doing some proper sizing of our environment so we don't find ourselves on an emergency all the time, which is what a lot of customers end up going through.

This is something I myself have experienced countless times during my career, and it's just people not talking to each other. They're all in the same building and they just can't go one floor up and a few cubes down the hall and just talk to the person that actually has an idea of what's coming down the pipe. Better communication and collaboration is really what's all in terms of capacity planning.

But when you mentioned something really interesting, like x gigabytes of storage, in the past you actually had to carve out very static portions of storage that would be dedicated to something.

And extended that would mean a data migration. That's all changing with underlying storage technologies where you no longer thick provision, like thick provisioning really doesn't make sense in 2018, in my opinion. Where you would dedicate a set of drives to do something, that's very 1999. Let's look forward and consider thin provisioning to be the rule rather than the exception.

And thin provisioning things so that you don't run into hard limits on things right off the bat. You don't have to spend too much time thinking about whether you have the capacity or not.

Mala: I'm a DBA who has to work with the SAN admin to decide a storage solution for a tier-one application. What approach should I take? What are the questions that I should ask?

Argenis: Well, number one, I'm very glad that the DBA is actually working with the SAN admins, because most of the time, they don't work together at all. The general conversation should be based on patterns and what the expectation on performance capacity is going to be like, not just capacity. When I say performance capacity, I mean not only IOPS, because IOPS are sort of like currency. This is something that I repeat very constantly out there when I try to explain to people what IOPS are. IOPS are IOs per second, but an IO can be of any size. I could be issuing an eight-megabyte IO. I could be issuing a 512-byte IO. They are very different in size because the payloads are very different.

So, the DBA should actually understand these things and know how the new tier-one application is going to impact or going to demand performance from that SAN, not just capacity. Here's a hard drive that's a terabyte, but that's not what you want. You actually want performance.

What performance do you actually want? I can give you very low latency-oriented performance, or I can give you very high bandwidth-oriented performance. They are very different workloads. Do you want something that mashes up the two? That is something that you should know. You should understand the patterns of your application are going to be. And if you don't have any idea what that application is going to look like, well then you should end up with the storage solution that is very flexible. That is actually going to allow you to handle both—at low-latency workload and at high-throughput workload, because you don't know what the demands for throughput are going to be like.

Let me give you an example. You're encompassing both transactional workloads and warehousing workloads. You're not defining any one of those when you say "tier one."

You can have both. You could have something that's actually very transactional oriented, sort of a financial application, think of it that way. But then on top of that, it will be ingesting thousands of rows of data per second or millions of rows of data per second. But in the end, you're going to have to do some reporting and some analysis on that workload and typically reporting analysis means warehousing, throughput-intensive, means don't care so much about the latency, cares more about the throughput of things. This is something that DBAs have to think about. They have to consider what are the implications of the workload that they're about to throw on top of a storage solution.

And this is where the SAN admin and the DBAs really have to work together and come up with a good analysis of what the demands of that application are going to be on the storage host. And whether that workload is going to fit on a given storage solution and some vendors are actually better than others at actually telling you whether the workload will fit. Because you can actually give those vendors some description of the workload and then the vendor can tell you if it's going to be able to handle it. So, you're going to need additional expansion. You're going to need potentially another set of devices, etc.

But it's important that you consider all these things. It's not just, oh I need ten gigs, oh I need ten terabytes. When I was telling you about storage testing and validation, I was talking about the actual shape of the workload in addition to the dataset itself. Those two things together are really what help you understand how a workload's going to run on top of a given storage solution. And if you don't have any data on that, you're really in a pickle. You really need to sit down and do some educated guesses as to how the application is going to behave. And if you don't know, you need to go talk to your business and your developers, because they're the ones who are actually writing the code that's going to hit your database.

The business may not be writing the code but they're given the developers the actual guidelines for how the applications going to behave like. So, go talk to everybody. If you're just waiting for requirements to hit your desk, you're not being proactive. You're not adding value. Adding value these days means actually traversing multiple layers of bureaucracy in companies. Becoming efficient and talking to multiple people at the same time.

Mala: How different would these questions be if the service were on a virtualization platform?

Argenis: That's always a great question. In my opinion one of the biggest problems with virtualization is that people don't understand, number one, how to virtualize workloads, what to consider when they're virtualizing workloads, and where they already have virtualized workloads running on a given infrastructure, what is going to be the impact of adding another workload on top of that. It's the exact same thing that I just described on the storage side, isn't it?

You don't know how to convert something that's going to fit around your storage device. It's an additional workload that's on top of something that's already running on the storage device is going to be. The virtualization problem is very similar in that sense, but people don't know what happens when you have to virtualize something. What happens when you have to throw a workload on top of it. Virtualization typically has this blender effect on the workloads. It's going to mash up all of these IOs coming from all of these different workloads together. So, the impact on the storage solution is going to be radically different than if you were placing one workload at a time on top of a physical kind of thing.

It's going to be a very different pattern that you're going to see from your storage devices. So, the truth is that there's nothing better than actually testing, replaying workloads on virtualized environments before you actually do the real migration. Just P2Ving a workload and throwing it on VMware or Hyper-V, hoping and crossing fingers that that's going to work is not good enough.

A lot of people just start praying. And guess what? That praying is not going to help you there at all. You might just get lucky, and that workload makes it and the solution that you architected works. But you have to do a lot of testing and there's no better testing than replaying. I don't know how many times I say this. You need to take your existing workload and replay it on whatever else it is that you're actually looking to put that workload on. You're going to have to make a big spend on these test environments. I'm very glad to be able to say you need to spend money on your test environment. If you're not testing, then you're running into a risk of fitting or converting a workload into something that is actually not suited for.

So, don't blame the people who are actually in charge of supporting that infrastructure if you didn't do your homework and you didn't spend your dollars on having a proper testing environment for that workload to be vetted as being able to fit into whatever you're trying to fit it on.

Mala: What's the role that documentation has played in some of the gigs you've had? How do you recommend DBAs work with documentation?

Argenis: The most valuable pieces of advice that I've seen on Twitter are the ones that relate to blogging. And the ones that relate to blogging relate directly to internal documentation. The number-one tell that you're doing it wrong is that you're saying the same things in meetings. If you're saying that and people are not reading it somewhere else, that's the first problem, why are people not getting that message, so that should be something to think about when it comes to documentation.

The other thing is, when documenting your environment, you shouldn't put that on a closed system that only you have access to. Make the infrastructure documentation available to everyone.

So that they can see what's out there, be completely transparent. Unless, of course, you're working on some security concerns and you're in a high-security environment where everything has to be kept under wraps. But you should have at least a wiki at your job. Maybe the wiki that comes with Jira, or even SharePoint has a wiki. There's a ton of different wiki-like pieces of software that you can use. You should be documenting stuff. If you are writing lengthy emails, you're doing it wrong. You should be uploading that stuff on wikis and blogs.

If you're having tough conversations, blog about them. Put them on a wiki. Have people refer to those things so that they know. They learn from all the things that you have gone through. You deliver value that way. Doing the same thing over and over again is wasting your time. You're going to be seen as a cost center, not a value center.

Think about that as DBAs. This is how you become better. This is how you educate others and this is how you actually adjust to the constant needs of your business. By making sure that you pay close attention to the things that are becoming repetitive and the need to be documented and the things that are causing you pain.

Mala: What are your personal favorite books, blogs, or other means of learning? How do you keep up?

Argenis: I don't keep up. I'm not a huge book reader. I do have a couple books that I'm going through. One that's really interesting, something Brent Ozar blogged about, is database reliability engineering. It's an interesting approach to managing databases.

And then since I'm doing a lot of work with VMs and SQL Server, I have a couple of older books from VMware folks that I'm reading. But, those are the only books that I've really paid attention to recently, I do spend a significant amount of time looking at blogs, though. I love the things that folks out there in the community come up with. I find the DBA From The Cold blog mesmerizing.

Side note. I think that containers are going to be it in the future. We're not going to talk about VMs as much as we're going to talk about containers, and that's my opinion even for SQL Server.

Twitter is essential. Even LinkedIn is pretty good in terms of disseminating information on what's out there. But I think Twitter is the way to see industry trends, and what people are talking about, and getting up to speed on blogs.

I don't know if you've been to the SQL Server community on Slack. There's a lot of people helping each other there. It's amazing. It makes my hair rise. It's so cool to see people helping each other like that. Free of charge, you just join. Start asking questions, and a whole bunch of other people will come help you. I think that's one of the coolest parts of the SQL Server community in general.

Mala: What are your recommended ways of stress management? How do you handle stress on the job?

Argenis: I don't handle that very well. Honestly, I try to go on vacation as much as I can. That's not always feasible from a financial point of view, of course, but I do like to try to disconnect for a week or two at least once a year.

Every once in a while, you have to unwind and do your own thing and not focus on the daily grind. But when I'm focused on the daily grind, I focus on the daily grind. Try to take a vacation and disconnect. That's important.

Mala: What's your style of interviewing a data professional? What do you look for in somebody you're looking to hire?

Argenis: You know people hate it when I say this, but I actually look for cultural traits rather than actual knowledge. I do focus a lot on the personality of the applicant, because that basically tells me if that person is going to be a good fit for the team.

Nobody likes the brainiac that doesn't contribute. Or when he contributes, it's basically all about him or herself. I like workers that are very communicative and like to collaborate. That naturally leads to a better team and better outcomes, rather than just asking what are the actual differences between the columnstore and columnstore archive compression—those are just details you can look up.

People can always go to the documentation and figure out that stuff. Now, should you have a systematic overview of how things work? Obviously, you wouldn't be a good professional if you didn't understand how things work but I don't focus on syntax. A lot of people ask if you can write how you would build an index. Give anyone five minutes and they can Google it. A lot of people focus on that kind of minutia, but to me, that really doesn't make sense these days. What you really should focus on is somebody that can actually think beyond that, and have a wider spectrum of vision, and get things accomplished yet be a good part of the team. Because you don't work by yourself. You're working with somebody else. Everyone has a teammate. Getting along with others, that's really the hardest part.

Mala: What are your contributions to the community, and why do you recommend people being part of the community?

Argenis: The things that I've done in the past are typically related to speaking engagements. I put in some time on the board of directors of PASS, spent a year there. I'm hoping that you'll agree with me that I gave to the community somehow.

The thing that makes the SQL Server and the Microsoft data platform community so cool is precisely that people are very involved.

I say that people are very open and willing to share. I took it upon myself to reiterate to people as frequently as I possibly can. If you take some knowledge to the grave with you, you suck. You need to share because people are going to be here on this planet after you're long gone.

Whatever you learn, whatever you share, is what defines you. If you keep things to yourself and you're not contributing and helping others, you are doing it wrong. We all get better by all of us contributing at the same time, all of us pulling everybody up. Rather than certain people pushing people up. Or you're trying to shine by yourself. It works in siloed places but the best way to give back is by sharing what you know. Even if that means that you're going to become uncomfortable doing it. Because that's actually a big issue for some people. They are struggling to get on stage and talking to others. Or maybe even opening their mouth at a local user group when they're having discussions or even spending additional time in addition to their regular jobs just trying to help others. Quite honestly, that's the thing that I hope that would have started earlier, helping others and getting involved in communities. That's a really joyful feeling.

Key Takeaways

- Always try to deliver value in terms of what you do, as opposed to just being seen as a cost center.

- The best way to perform storage evaluation and testing is to just replay the database workload on top of a restored version of the database.

- No company's going to be 100% as successful as they could be by focusing on a portfolio from a single company as a storage vendor.

- Too often, DBAs are waiting for requirements to come in, and that's actually the wrong approach. The right approach is talk to everyone and understand what's actually coming.

- Thick provisioning really doesn't make sense in 2018. Where you dedicate a set of drives to do something, that's very 1999. Let's look forward and consider thin provisioning to be the rule rather than exception.

- IOPS are sort of like currency. This is something that I repeat very constantly when I try to explain to people what IOPS are. IOPS are just IOs per second, but an IO can be of any size. I could be issuing an 8-megabyte IO, I could be issuing a 512-byte IO. They are very different in size because the payloads are very different. The DBA should actually understand these things and know how the new tier-one application is going to impact, or going to demand performance from that SAN, not just capacity.

- There's nothing better than actually testing or replaying workloads on virtualized environments before you actually do the real migration. It's not good enough to just P2V a workload and throw it on VMware or Hyper-V, hoping and crossing fingers that that's going to work.

- You need to spend money on your test environment. If you're not testing, then you're running into a risk of fitting or converting a workload into something that is actually not suited.

- If you're having tough conversations, blog about them, put them on a wiki, have people refer to those things so that they learn from all the things that you have gone through, and you're not having conversations over and over again. You deliver value that way.

- If you take knowledge to the grave with you, you suck. You need to share because people are going to be here on this planet after you're long gone.

Kirsten Benzel

Senior Database Engineer, SurveyMonkey

Kirsten Benzel *is a mischievous senior database engineer who has been working for SurveyMonkey since 2014. Prior, she enjoyed the deserts of Arizona while working on-site for GoDaddy, and then found the Bay Area also to her liking while she worked remotely for several years. She became fascinated with computers at an early age. When she's not gaming, you can usually find her gleefully tuning a query or wandering a beach looking for sea glass. Her career in information technology began when she was unable to find employment as a town philosopher after earning her BA in Philosophy from Northern Arizona University. She became visible in the SQL community while twice administrating Argenis Without Borders, a fundraiser for Doctors Without Borders that ended during the annual PASS Summit. She can be reached on Twitter at @cybersnark.*

Mala Mahadevan: Describe your journey into the data profession.

Kirsten Benzel: I grew up with computers and for that I'm lucky. I remember eight-inch floppies and the days when a file name could only be eight characters! It's because of this early exposure that later on as an adult I was able to successfully launch into IT. I think early, hands-on access to computers is absolutely critical in removing the intimidation factor that so often steers young people away from a career in IT.

© Malathi Mahadevan 2018

M. Mahadevan, *Data Professionals at Work*,

https://doi.org/10.1007/978-1-4842-3967-4_20

After two semesters in college, I chose my major by reviewing the courses I had completed and identifying correlations between the ones that I enjoyed—all very data-driven. That led me to a BA in Philosophy with some HTML, CSS, and C# on the side. I admit to being a terrible "traditional" programmer even then, and the advice I was given early on wasn't helpful in finding a better path forward. If you had asked the college version of me what my plans were, I would have replied that I knew I needed to learn programming languages like C, C++, SQL, Python, and Perl. I had no idea which way to go nor any idea what that list even had in common! My advice to more seasoned advisors would be to learn what your protégé has tried, then learn their strengths, and from there recommend just one or two similar technologies that could help them into a general area they might enjoy.

Mala: Describe a few things you wish you knew when you started your career that you know now and would recommend newcomers to this line of work know?

Kirsten: Luck plays a big role. I started out as a call center representative knowing exactly nothing and worked hard over the span of five years to become a database developer. It's because of luck that I landed in a company that promoted internal hiring and opened doors. But it's hard work and the ability to soak up information that ultimately got me the role.

No one gets into database work on purpose, and it's a shame! Colleges teach rudimentary SQL usually as some auxiliary to something else, and that's it. There's a reason you see a slew of Accidental DBA classes in the wild. I wish I had known SQL was "a thing." It's really up to us—industry professionals—to shoulder the task of educating colleges and parents on its very existence and necessity.

I've always loved Excel, data, making correlations, and drawing conclusions. We're doing a disservice to young people just like me who are blind to this as even being a career option. You'll often hear about the lack of women in STEM [science, technology, engineering, mathematics], and I think these two issues have the same root cause: a lack of early exposure. Have your child build a website, learn what a web server is, and spin up a little MySQL database on the back end as their class science project. Make a database of frogs, or horses, or dinosaurs, and display them on a basic site. It might seem intimidating in print for me to just haul off and say that, but we literally have a world of tutorials at our fingertips with a single click.

When I'm talking to someone who already has an interest in database work, I always recommend "the book"—T-SQL Fundamentals by Itzik Ben-Gan [Microsoft Press, 2016]. This is my way of passing along some of the best advice that I was given by our senior DBA, Pete, who said, "Read anything by Itzik Ben-Gan." That advice is still rock solid even now, ten years later.

Mala: What was analytics/data visualization like five years ago? What has changed, and why is it such an "in" thing now?

Kirsten: Five years ago, we didn't have the raw data that we have today. I realize this might seem like I'm stating the obvious, but when you consider the difficulty of signal to noise and then trying to visualize it and explain it to an audience who has a need to know, it can be daunting. We joke in the industry about the term big data and what it means. I like the definition of a course on Amazon Redshift that I just completed. "Big data is high-volume, high-velocity, and/or high variety information assets that demand cost-effective, innovative forms of information processing that enable enhanced insight, decision making, and process automation." Data visualization is not easy, and as our data sets grow and change, so will the tools advance and the need for experts grow. It's a promising area, especially if coupled with a solid foundation in statistics.

Mala: What are some ideas one can present to management as a BI person as far as analytics, big data, and buzzwords like that go?

Kirsten: Buzzwords are a personal pet peeve of mine! Every time I hear the word ask used as a noun, I cringe. At one point at work, we were renaming all our conference rooms and I suggested we use buzzwords and phrases. Take It Offline, All Hands-on Deck, On Your Plate, Outside the Box, Close the Loop, The Next Level, Core Competency, Window of Opportunity, Low-Hanging Fruit. I made a list of thirty-plus! My recommendations weren't considered, but boy did we have a lot of fun making the list.

I think imprecise language is annoying at best and damaging at worst. Let me give you a real-world example. A common metric that many companies use is engagement. On the surface, this seems like an obvious thing. It's how much users engage with the product. But when you're trying to measure it and present, say, a line chart of "engagement," what exactly does that look like? You have to go much, much further and define the metrics you want to measure, whether that's logins, page visits, page clicks, page hovers, or purchases. To tell my VP that "engagement fell off" is alarming! And much less meaningful than perhaps, "Starting seven days ago, page clicks to the English version of the billing page fell off by fourteen percent." I don't personally use buzzwords, and I view speaking properly, exactly, and directly as an exercise in ethics. You should read La Parole by Georges Gusdorf [Northwestern University Press, 1979] if language and ethics interests you. Precision of language is key, as is adding as much relevant context to a statement as possible, which leads into the next section on storytelling.

Mala: Some people think storytelling means weaving stories around data. Data should just present facts. What is your take?

Kirsten: I've never heard that data should be presented as a story. My first inclination when I hear this is to ask why? In my current and previous roles, it's been my job to present accurate, timely, raw data. Just facts. Maybe the story-thing is relevant to analysts who have to answer why.

In my opinion, if I do my job right, then it should make the "intermediary why" easier. What do I mean by this? Tell me why orders for appletinis dropped off starting last Tuesday. Well, if I show you an unsorted list of our martinis, you might just smile and nod. If instead I show you the same list but sorted by the primary flavor with the two flavors that are out of stock crossed out at the top, you can now very easily know why I've brought this to your attention. The "intermediary why of "Why is the apple flavor out of stock?" directly relates to the why of "Why did orders drop off?" It's a crude example, but you get the gist of it.

The quality of how I display data to our analysts is, I think, just as important as how fast and how often it's accessible. Smart BI highlights anomalies. It's perhaps not my job to tell the story, but I can try to point the team in the direction of an interesting one if I'm doing it right.

Mala: What is your experience with agile methodologies and business intelligence?

Kirsten: I've used agile at both companies that I've worked in IT for—GoDaddy and SurveyMonkey. I love the idea of sprints, and when managed properly, they're incredibly useful. On the other hand, two of the most difficult things in IT are estimating the time it's going to take to do something and naming things. Go ahead and laugh at the naming thing, but you'll see!

My general rule is to make a time estimate guess and then triple it. If I think something will take me five minutes, it will probably take fifteen. A one-day project should be blocked off as three. If I may slice and dice a few clichés: over-estimate and just deliver.

I think it's harder for a BI team to use agile than other branches of the organization. The reason being is that BI or data teams are in thrall to most other teams. We can't create a report until operations deploy the feature. We can't analyze the A/B test until the data comes in. Meanwhile, we're being asked questions from project managers and marketing, and all the while infrastructure is trying to optimize and improve processes, and code, and reports under our feet. When you try to put all those moving parts into neat, tidy, little two-week chunks, you need to have a top-notch team or be excruciatingly patient with scope creep.

Mala: Describe your experience with cloud adoption.

Kirsten: The cloud is another one of those buzzwords that really ruffles my feathers. Someone funnier than me said that the cloud is just "somebody else's computer" and I prefer that definition. There's nothing spooky about it.

You're literally just loading your data somewhere else and processing it there instead of here, on premises.

My experience is limited but that's changing fast. As I said earlier, I just finished a course on AWS Redshift. The cloud is fantastic for distributed processing. What's that? Well, I really like analogies. If you can imagine a basket full of apples that need to be peeled, and then imagine one person peeling one apple at a time, you can see where that might be suboptimal—a favorite term at work! Instead, give ten people each a handful of apples and yell, "Go!" That's distributed processing in a nutshell.

If you group your data into files—yes, just like the .txt files you see on a consumer grade computer—split out by month, then you can have many threads—thread equals apple peeler—chew through your data all at once in order to answer a date-related question. It's much faster, but it means you must choose carefully how you physically slice your data.

Distributed processing is pretty neat, but there's a lot of factors to consider before you choose the cloud as your poison of choice. The how and why of that fall well outside this chat, but hopefully, I've at the very least generated some moderate interest in using cloud computing.

Mala: What are some of the common data quality issues when dealing with data? What can be done to avoid them or to mitigate their impact?

Kirsten: End users and data have something in common: if they can, they will. No, really. The things I have seen people do and the data that's gone by on my screen just boggles the mind. When you're wearing your QA hat, you must try and do every possible thing that an end user might do. Yes, someone really will paste the entire contents of a book into the text field that dev didn't put a limit on so you better be ready. Because of my experience—or maybe my lack of imagination?—I really prefer the whitelist approach over the blacklist approach when it comes to dealing with data integrity.

When I'm designing an ETL or a new database, I choose the correct datatypes and data lengths based on what should be allowed. This is called the whitelist approach. I'm giving the okay or "whitelisting" allowed behaviors and values. Why this approach? If you take the blacklist approach and only define what isn't allowed, you will be forever lost in the Dungeon of Guess What Bad Data Came in Today. Roll fortitude. Your life will be an endless list of rules that you have to keep updating every time you encounter a new violation. And that's just suboptimal.

What does this actually look like? Some examples: no leading or trailing spaces. An email address must have an at sign and a dot to be valid. An IP address must have four dots or at least two colons to be valid with a maximum length of thirty-nine for a fully enumerated IPv6 address. Use NVARCHAR for anything an end user is going to enter so you don't HORK the data. Use Python to do string matching because RegEx options in T-SQL are scarce. Use the most current time type that you can because Microsoft improves them every time. Right now, that's datetime2. And last but definitely not least, provision your servers to use UTC as the default and store your timestamps as UTC. Trust me on that last one.

Mala: What are some of your favorite tools and techniques?

Kirsten: I did a presentation on how to go from new hire to deployment complete in one day or less. One of my tricks is to have three programs installed: Softerra LDAP browser, Sublime Text, and Redgate SQL Compare.

Why do you need an LDAP browser? Because when you're first starting out and even after you've been doing the job for a decade, you're going to get asked Active Directory questions. It's going to make your IT department so much happier if you tell them what group you need to be added to after you've confirmed that it actually exists and you've spelled it right. Plus, "Why can't I see this database? I used to have it!" is at a minimum a weekly question you'll get from users, and you'll want a quick way to verify or compare AD groups.

Notepad is evil. Use it for notes that you don't care about. The big kids use Sublime Text! It does so much. It's color-coded, and you can compare documents side by side, to name but a few reasons why. Also, did you know that in Notepad [the default text editor installed on a Windows machine] when it's doing a find-and-replace it returns to the top of the document after each iteration? So suboptimal. Although to really level up, you'll want to become a Vi/Vim master.

Last, grab something to compare database objects. Choose a program that doesn't lock the schema as it works. From day one, you're going to be comparing database objects and occasionally the data in small lookup tables, so get one of these, and get it quickly. I like Redgate Compare.

Mala: What is the role of documentation in being a good BI/analytics person?

Kirsten: A wise person once told me that "as soon as you write it down, it's obsolete." There's some brutal truth to this, but it doesn't mean you shouldn't document anything.

Documentation is critical to the success of your team. If you skip to the question about work/life balance, you'll see how important camaraderie is, which reflects the importance of documentation. In an ideal world, tribal knowledge would be a thing of the past. I think perhaps it's best to shoot for a happy medium. If you can hand a new hire your documentation and have it get them somewhere around one-third of the way to "up to speed," then you're doing pretty good! At a minimum, you should keep a data dictionary for all the metrics and slang and the endless, company-specific acronyms. Network diagrams never hurt. Common workflows and ETLs, and of course, it's nice to have an onboarding where-to-find-it document.

I just recently started using Alation, and I couldn't love it more. Not only does it automatically pull in database schemas and stored procedures but you can easily create really slick-looking documents and link them to code snippets and database objects. Sure, this sounds like a sales pitch, but maybe it is? I've always personally struggled with documentation as a chore, but I'm finding I actually look forward to using Alation. That's high praise.

Some people hoard information either by accident or as a misguided job security technique. Keep in mind that if they cannot replace you, they cannot promote you. If no one else can do your job because there's been zero knowledge transfer, then you'll be on the hook 24/7/365, and that's not healthy for you or the company.

Mala: What are your favorite books, blogs, or other means of learning?

Kirsten: I'm a huge fan of RSS feeds for keeping up with technology. Another good resource is the free SQLSaturday that PASS sponsors. If you don't know what a SQLSaturday is, you're missing out. These events are where the industry experts—the movers and shakers, if you will—present talks and classes, every Saturday, to help promote community and to help educate attendees. Find your local chapter.

Last, I like Twitter as a means of keeping my thumb on the pulse of the industry. I love how it's an easy way to keep up with everything at once, and I consequently hate how sometimes it's like trying to drink from a firehose. Learn to create lists and keep the people you follow in tidy groups so you don't get overwhelmed or succumb to the brutal context switching that your main feed consists of.

Mala: What are your recommended ways of stress management and developing healthy work/life balance? How do you handle long grinds on the job? How do you work long hours that are boring and not challenging while still keeping motivated?

Kirsten: I struggle with this because I'm a work "sprinter" and not a "marathoner." If you're like me, you go on four-to-six-hour benders and never take the headphones off or leave your chair. Well, the facts are in, and this is disastrous for your health both physical and mental. I've just recently created alerts on my phone to remind me to stand up and stretch hourly and to eat something every three hours. I'm a recent convert to Bulletproof coffee and I'm a big fan of Vital Proteins Collagen Creamer and organic maple syrup in lieu of the old traditional sugar and creamer. Add a dash of cinnamon, and you've got delicious brain food right in your hand.

Sleep is tough, too. When you're under deadline pressure, it's almost impossible to rest sometimes. Deep breathing gets me to sleep if nothing else works. For those wicked on-call rotations, I'd recommend setting up a buddy system—this is where documentation is key—so once you've had two or three bad nights, you can hand over the pager to a buddy so you can recover. Similarly, keep an eye on your teammates, yank the pager from them, and cover if they've been suffering for a few days and are too tired to notice. It's a give and take.

Being constantly connected is probably the worst part of working in tech. I cannot count the number of times I get an envious response when I tell people that I can work from home, because what they don't realize is that the laptop is not freedom, it's a tether. It's up to you to set boundaries, and I really struggle with this, too. If I wake up at three a.m., bored, I'll often hop on and work without the usual interruptions. It's up to me to then clock out at a reasonable time and not overdo it. Otherwise, my boss jokingly threatens to yank my AD permissions!

Another thing that's relatively new is the concept of unlimited vacation. It's one of those deceptive things that sounds good on the surface but sometimes has hidden teeth. With the introduction of unlimited vacation, the burden of how many days off is appropriate has been moved from the company to the employee. Now, if you are perceived to take off too much time, it can have a negative impact on how your work ethic is regarded. Again, managing this in a healthy way is up to you. You must take time off to disconnect and recuperate, or even to take care of family needs, and the best balance is going to be different for each person. Don't feel guilty for taking personal time to recharge.

I can't finish a section on work/life balance without addressing the B word—burnout. It's real. It's prolific. And if you don't set boundaries, it's going to bite you really hard and really fast. Once you're burned out, I honestly don't know how you recover, so try not to go there. I have a slide saved on my phone with the six causes of burnout by Cate Huston:

1. Lack of control

2. Insufficient reward

3. Lack of community

4. Absence of fairness

5. Conflict in values

6. Work overload

I'm lucky enough to work in a great company and on a superb team that makes one through five a non-issue for me. Work overload is tougher. As database professionals, there is always more work to do. In this job, you never, ever clear the queue. And somehow you each individually have to make peace with that. Again, disconnecting is key, even if you choose to just sit at home and stare at the wall! If you feel burnout creeping in, try to reach out. If you cannot, it may be time to look for a place with values that more closely mirror your own. Remember, not all matches work out and that's okay.

One last thing. It's not about how you fail, it's how you recover from failure. Be transparent, take ownership, fix it quickly, and learn from it.

Mala: Describe your style of interviewing a data professional. What do you look for, and what are some examples of questions you ask?

Kirsten: I conduct two types of interviews—technical and non-technical. The technical interview is a list of technical questions that we as a team have created for the open role. My non-technical interview, on the other hand, is an hour of talk. I'll share my top-two favorite questions to ask.

Pretty sure I stole this one from Brent Ozar. I like to ask the interviewee to tell me about something they've recently done, a project, that they're proud of. It doesn't even have to be SQL related. I'm looking for that spark, and to start up a dialog where they genuinely want to share with me something cool. Interestingly enough, you cannot fail this question, and it's led to some really quite interesting conversations that I've enjoyed.

Second, this one I borrowed from a TED Talk by Adam Grant on givers and takers—highly recommended! I ask them to list three to five people whose career they have influenced. The ideal answer will be a handful of people in positions equal or lower to them, who they have mentored.

Mala: Can you narrate a funny or an interesting story to share with our readers?

Kirsten: A man lived in a poor area in a small home with a dirt yard. Every morning he'd watch from the window as a neighbor dumped trash into his yard. He never said anything and quietly cleaned it up, day after day. One day, the neighbor did not appear with the trash. The man hurried next door to make sure his neighbor was okay.

Even if you have to keep cleaning up messes, remember what's important.

Key Takeaways

- Precision of language is key, as is adding as much relevant context to a statement as possible is important.

- Smart BI highlights anomalies. It's perhaps not my job to tell the story, but I can try to point the team in the direction of an interesting one if I'm doing it right.

- End users and data have something in common: if they can, they will.

- Keep in mind that if they cannot replace you, they cannot promote you.

- It's not about how you fail, it's how you recover from failure. Be transparent, take ownership, fix it quickly, and learn from it.

Favorite tools: Softerra LDAP browser, Sublime Text, Redgate SQL Compare, Alation

Recommended books: *T-SQL Fundamentals* by Itzik Ben-Gan, *La Parole* by Georges Gusdorf

Tracy Boggiano
Database Superhero, Broadvine

Tracy Boggiano is the Database Superhero for Broadvine. She has spent over 20 years in IT and has used SQL Server since 1999. Tracy covers all aspects of administration and deals heavily with performance tuning, high availability, and disaster recovery.

Tracy is co-organizer of the TriPASS Advanced DBA Group. She is also the founder of WeSpeakLinux. com. She currently leads getting Linux content into the various PASS Virtual Groups because she is super excited about SQL on Linux. Additionally, she volunteers to mentor people on SpeakingMentors. com. Before Tracy worked full-time as a DBA, she was formally a developer and a system administrator.

Tracy wanted to work on computers since she was six years old, and she tinkered with databases in middle school and high school to keep her sports card collection organized. She has volunteered through the North Carolina Guardian ad Litem program since 2003, advocating for abused and neglected foster children in the court system. Volunteering with foster children is her passion outside of SQL Server and her favorite job. Tracy tweets at @TracyBoggiano and blogs at Database Superhero (http://databasesuperhero.com).

Mala Mahadevan: Describe your journey into the data profession.

© Malathi Mahadevan 2018
M. Mahadevan, *Data Professionals at Work,*
https://doi.org/10.1007/978-1-4842-3967-4_21

Tracy Boggiano: I got a computer in seventh grade, and I got started with databases at the time because I had a sports card collection that was very important to me. I kept up with it in a program called MyDatabases, and then I didn't touch databases for a few years after that because I didn't have a computer. I jumped back into computers at eighteen, as soon as I got out of high school. I managed to get a job as a computer operator, where I really was like a jack-of-all-trades. I ended up with an Access project that was a reporting front end to a SQL Server. We had links to all the SQL Server tables and a lot of reports set up.

I developed a lot of reports. And then eventually I turned into a web developer, and I actually had to do something with SQL Server. I figured out that the databases on SQL Server were not being backed up and that they didn't have any other maintenance tasks happening. So, I set up maintenance plans for indexes, backups, and integrity checks. At that point, I was studying for my MCSE [Microsoft Certified Solutions Expert] certification, and I decided after that to study and learn more about SQL Server and go for my MCDBA [Microsoft Certified Database Administrator] in SQL Server 7. I learned a whole lot of things about SQL Server by doing that, and I set up a lot of processes in that company. However, I stayed a web developer for a few more years before going through layoffs. Afterward, I took a full-time DBA job, and I've been a DBA for almost twenty years now.

Mala: Describe a few things you wish you knew when you started your career that you know now.

Tracy: One thing I did not know was that the community that we have—SQL PASS—existed. I didn't really know about it until I was doing DBA work for seven or eight years and learned about it at a conference. Once I started doing that and started reading blogs and books, I came across SQLskills first after attending one of their trainings. Through that I was exposed to Paul [Randal] and Kimberly [Tripp]'s work and then branched out from there. I guess just knowing that there are so many resources out there—blogs, SQLSaturday, and Twitter.... In my opinion, those things can help you so much more than picking up a book and trying to learn on your own.

After that, a couple years ago, I actually started getting involved in the community. Once you become involved, you really start meeting everyone and learn how much the community is there for you when you become a part of it. You see a little bit of it when you're using the community from afar, but then, once you're in the community so-to-speak, you discover there's so much more.

Mala: Do you have any experience with agile methodologies as a DBA? Do you wish to say anything in that regard?

Tracy: We use agile at our current shop. I just kind of pray when we have our deploys every month, because there are a lot of changes that happen on the second Tuesday of the month. But we have good database developers that make sure that good code gets deployed. I think the most important thing—if you're in an agile shop and you're just the DBA doing the admin part of things—is to make sure you are collecting good performance metrics so that when code is deployed, you can easily see if something has changed and track it down. All of a sudden, the CPUs spiking to one hundred percent, or we're seeing a lot of blocking, or our PLE has dropped to nothing. If you don't have any metrics to know what's happening when you're in an agile shop and things are changing rapidly, you are kind of sunk.

Mala: Are there any tools and techniques that you really like?

Tracy: Yes. We are currently using an open source project called Telgraf that gives us a lot of graphs set up in Grafana. I have a blog series about it. It allows us to see everything in a graphical format that's pretty interactive, and we can make our own graphs for everything. Our shop is pretty unique. So, we really can't use the third-party tools as well as we'd like to be able to.

We also take advantage of using sp_whoisActive to capture everything into a table every minute so we can find the long-running processes. We love using Query Store. We upgraded to 2017 recently. We're taking advantage of the auto plan correction. And anything that I can make happen automatically to make it so that I am not doing it manually, I love to set up.

Mala: What sort of struggles have you faced with managers, management, or the business side of things, and what's your recommended approach to handling that? You don't have to get job specific, you can keep it generic to the extent you want.

Tracy: Management-wise, the biggest struggles I've ever had is when you end up with a manager that is not as familiar with your technology as you are. One time, I had a manager that was more familiar with mainframes than they were with any Microsoft technology. So, trying to explain how something could get deployed to one Windows machine out on the manufacturing floor differently than all the other machines, was not something that the manager could grasp, because the mainframe was just one deploy.

Dealing with other teams I think depends on how well a relationship you develop with those teams. I like the advice I heard from others to bribe other teams with food and beer. I tend to be one of those people that can get along with anybody. It's one of my superpowers. I've had the occasion where I've had one bad apple at one company who soured my relationship with a few teams. So, I would send my teammate over to those teams and have him engage with them when I needed something from them. Certainly an unfortunate situation, but management did not step in to help correct.

Another strategy was to let my manager handle those team engagements if it came to a point that it was difficult enough to where they wouldn't deal with me at all. I've had situations with SAN admins and network admins where you had to prove to them that it wasn't the database before they would even look at anything. Come on, we're all the same team here. We all want the servers to run fast. We all want the users to be happy. That's all I care about. I don't particularly care where the problem is. I want the servers running fast. I want the application running. I want the customer happy. I don't care where the problem lies. I don't care what caused the problem. All I care about is getting it fixed.

Mala: What are some industry trends in this line of work that you're really excited about?

Tracy: I like the automation trends, where some of the things that we're doing day-to-day take care of themselves. I do like the auto plan correction in Query Store, because that opens up time for us to do other things.

My time is heavily committed, so for me to not have to go in and click the button to auto correct the plan or to recompile plans, even though I have a process that automatically recompiles plans, just that one fact that we won't have to do that is kind of nice. Setting up things in the cloud is nice because now you don't have to worry about the hardware as much.

There's still going to be a need for us as DBAs because you're still going to need to plan things out. For high availability and disaster recovery, for example, you're still going to need to be able to make sure that you're able to meet your service-level agreements. For database development, somebody's still got to write the code efficiently. No one can do that for you effectively but a DBA. Entity framework is not one of my favorite things. Somebody actually has to know how to use it correctly in order for it to work well on a system, and somebody still must properly design the database itself, or it's not going to perform well.

You can't just slap a table in there and not design it correctly and think it performs well. I don't see DBAs going away with some of the things that are being automated. It's just giving us the ability to concentrate on other things.

Mala: What's the role of documentation in your job? How important do you think that is for a DBA?

Tracy: I think it's very important. I believe we don't manage or leverage it as well as we should. This is something we've been trying to change where I work, and we've played with different formats over the last year. I just think any process that you write needs to be documented. When you have a team of five or six DBAs, you tend to have one person that wrote a process and that is the person that understands it the best. It needs to be documented so the other five DBAs on the team can look at it when they are on call or you are out, and they have something they can go feel comfortable enough to be able to troubleshoot or use when you are not around. I think documentation is a good way to make sure any gaps in your team are bridged because everything is out there and available for everybody else to look at. It enables them to effectively take over when necessary.

Mala: What are the things you've seen as worst practices and you recommend that DBAs should watch out for?

Tracy: Terrible table design. In one position I had, they had a consulting company come in before I started there. Every table had only one primary key and it was a unique identifier because we're using replication on eight out of one hundred tables. So, they had decided that they needed it everywhere. They didn't even specify it as a rowguid column. So, then the tables that they used for replication, which had two unique identifiers because replication automatically added another unique identifier as a unique key. Additionally, they didn't do any indexing. I came in and re-architected everything in that database. I added indexes and made things go from five minutes to two seconds.

I also think that as a DBA, you need to practice restoring your databases. I had a situation one time where a system table that stays in memory was corrupt, and we did not find out until we did maintenance and rebooted the server. I believe you should have an automated process that restores your databases to another system to make sure your restorers actually work. Also record how long it takes. That way, you can go in and make sure you can meet your SLAs and run CHECKDB. I have more of an infrastructure DBA type role. So those are the things I am more concerned about—making sure my systems are up, and if something goes wrong, I am able to undo the wrong as quickly as possible.

Mala: Describe your experience with cloud adoptions.

Tracy: We use AWS [Amazon Web Services]. We have a few EC2 instances out there. We have not really adopted any RDS as an AWS, or any Azure SQL databases yet. So, my cloud experience is limited right now. I recently attended the SQLskills training on Azure. I think cloud has the potential for some things. It is not too different than on-prem besides you do not have to worry about the hardware. So, I am looking forward to exploring more of the cloud, personally.

Mala: What are your favorite books, blogs, and other means of learning? How do you keep up?

Tracy: I'm going to be a little biased and pitch a book that I helped write a chapter in. SQL *Server2017 Administration Inside Out* [by William Assaf, et al., Microsoft Press, 2018].

I collaborated in the writing of one of the chapters, although there was some editing to the content post-work. Helping to write it was really fun. I do love reading Paul and Kimberly's material, and their whole team's blog, because they do deep-dive type content of SQL Server. I like the SQLPerformance blog. I love Pluralsight. There's a lot of resources on Pluralsight. I use Curated

SQL because he links to a lot of different blogs on there. And then from there, I'm active on Twitter for everything else. I follow people I find interesting and knowledgeable, and keep up with many things from there. It takes a lot of time to keep up with everything.

Mala: Are there any favorite conferences or training that you like to attend?

Tracy: I love going to SQLSaturday. I'm a little bit of an addict on that end. I do love going to PASS Summit. It's a great networking and training event. Those are my two favorites.

Mala: What are your recommended ways of stress management and developing a healthy work/life balance?

Tracy: Not sure I really have many stress management and work/health/life balance techniques. Outside of work, I volunteer with foster kids. So, it is like a mini part-time job. But I do try to make sure I get some exercise. I try to go for a thirty-minute walk every day. I guess this may go into stress management. I plan every fifteen minutes of the day on a calendar, and I try to stick to that schedule. That helps me ensure I'm taking care of stuff that's not work-related. So, I've got every time block scheduled, even free time is scheduled, that way I know the time I have committed and the time I have open to do personal things or relax. I try to stick to that schedule as much as possible. If I see that I've got too many things on my schedule, I just have to say no to something. With all my community involvement in SQL and my volunteer work, right now there's not a lot of free time on my calendar. I am working on making that better.

My calendar scared one of my coworkers the other day when he looked at it. It is because I literally have everything scheduled. Every time block.

Mala: What's your style of interviewing a data professional? What do you look for in somebody you'd like to hire?

Tracy: Usually when I'm doing the interviews, I try to look more at how the person's personality will fit into the team, because nothing disrupts a team more than someone that does not fit well into the culture and dynamic of the team. They usually already go through a partial technical interview before I get to interview them. The manager will also ask technical questions during the interview process. I will ask some technical questions, but overall I look for somebody that has some experience dealing with developers and being able to politely tell them, "No. We've got to do it this way because of XYZ reason for performance or a design reason."

One of my favorite questions to ask is what kind of impact that person wants to make in the first three months, six months, and year, because I like to know that the person wants to come in and jump into something rather than just sit back and watch for the first little bit. I like to feel that they actually want to come in and feel motivated to get something done.

Even if they are not ready to make changes yet because they don't know the full extent of our environment, I'd still like to know that they're going to come in and want to make changes and make suggestions. So, knowing that they're able to do that, and that they're a good fit for our current team.

Mala: What are your contributions to the community, and why do you recommend that people be involved with the community?

Tracy: I try to do a lot of blogging. I'm trying to blog a few times a month. Especially about things that I'm doing that are kind of unique, or things that I'm learning or helping people with, because I get questions from the community now that I'm a part of the community. I think that helps to contribute to other people's learning if they have access to good reading material. And if they can pick up something from my site and learn about it, that's great.

I've just relatively recently started getting heavily into the community within the last three years. I've presented it over thirty SQLSaturdays. I think presenting is a great way to show people what you know and get to know people. By doing that, I've built up a little bit of exposure, which has led to, for example, this interview opportunity. Before that, nobody had that awareness about my expertise within the industry. It helps other people as well because everybody has a different view on things. They can learn something from your experiences versus someone else's experiences, because everybody's using everything differently.

I got to present at Summit last year, which I thought was really amazing and totally frightening at the same time. I got up on the stage and watched two hundred people walk into the room one by one.

That was my first time on a mic. I got there right when the other guy ended, so I was ready. And I thought, "Now I've got ten minutes of silence." But once I started talking, it was fine. I cracked my first couple of jokes about myself and went on about my way.

I helped with organizing a SQLSaturday by trying to broadcast that we were having it and getting people to sign up for it. I've been leading getting the Linux content into the various PASS virtual groups since July because I think Linux is going to be amazing. I think they announced on Twitter yesterday that forty percent of Azure is running on Linux, and I think ten percent of the SQL sales are Linux. So, Linux is going to be the way to go. Those of us that are only comfortable with Windows are probably going to need to learn some Linux.

I also organized the PASS Linux marathon that they had in December, and I won the outstanding PASS volunteer award in November for organizing the PASS Linux marathon.

I thought that was cool. I run a local user group for advanced DBA topics because we've got a few nerdy people around here, so we get to play with all the "cool kids," because we get all the people that do the level 300, 400, 500 sessions to present to us. Usually remotely. That's neat.

I started a website called WeSpeakLinux.com to get some more of the basic Linux stuff to the Windows people. So that's a neat thing to start. I joined SpeakingMentors.com as well to try to get people speaking into the community. So, I've actually picked up a couple people I'm mentoring there. I've had the opportunity of mentoring people at my workplace and previous workplace as well. I just think doing all of these things helps you become a better person and helps the community at large because everyone's got something to learn from someone else.

Mala: One last question, if you had one superpower, what would it be and why?

Tracy: I want to be able to teleport myself or like what they did on *Star Trek*, where you get on the beam and then you're somewhere else. And that's because I spend too much time in my car driving around the state of North Carolina visiting foster kids.

And it would help me get to more SQLSaturdays too. So, it would help the community as well. Then I wouldn't be driving all up and down the east coast.

I flew to Chicago on Wednesday for the SQLskills training, and then I flew into Pittsburgh at midnight, drove an hour over to Wheeling, West Virginia for SQLSaturday. I left the event at three o'clock p.m. and drove back home. I got home at midnight, and then at ten the next morning, at work for maintenance. If I could teleport, things would have been a little bit easier.

Key Takeaways

- The most important thing, if you're in an agile shop and you're just the DBA doing the admin part of things, is to make sure you're collecting some good performance metrics so that when code is deployed, you can easily see if something's changed and track it down.

- I want the customer happy. I don't care where the problem lies. I don't care what caused the problem. All I care about is getting it fixed.

Favorite books: *SQL Server 2017 Administration Inside Out* by William Assaf, et al.

Recommended training/conferences: PASS Summit, SQLskills, Pluralsight

Blogs to follow: SQLskills.com, SQLPerformance.com

Dave Walden

Cloud Architect

Dave Walden *has over 20 years of experience in large, high-performance, mission-critical environments. While his primary focus is the data platform, Dave has deep expertise in other fields, including the cloud (private/public/hybrid), storage, virtualization, compute, and networking. He also has deep experience in the big data space, having worked in technologies such as Hadoop, data lakes, Cosmo DB, and MongoDB.*

Dave is an international speaker on Cloud and SQL Server–related topics, most recently SQL Server on Linux and containers. He can be found on Twitter at @DBADAveKC and blogs at dbadave.com.

Mala Mahadevan: Describe your journey into the data profession.

Dave Walden: I kind of got into it by accident. I was primarily an infrastructure person—you know, hardware, software, operating systems. I got started in IT when I was really young. I was fixing computers when I was about ten, and I started doing some custom development when I was around twelve. I was making money on fixing neighbors computers, learning things, and doing a lot of stuff on bulletin boards. I fell into the whole data thing by virtue of one of my custom app development requests. I was probably fifteen.

M. Mahadevan, *Data Professionals at Work*,
https://doi.org/10.1007/978-1-4842-3967-4_22

I started working with Access as a database. I got the whole relational theory, and started reading as much as I could about relational database management systems. There wasn't a lot of data out there because this was probably 1996. I just started reading everything I possibly could. I just lucked into a few of those types of engagements and eventually got into a job with a company called GoldMine Software/Front Range Solutions. It was right at the time they were being acquired by a private equity group. At the time, they were one of the largest ISVs [independent *software vendors*] of SQL Server, so they had a lot of priority access, really good talent, and a lot of information, so I was able to learn quite a bit. And it took off from there.

Mala: How did you end up a cloud architect?

Dave: To me, the cloud was just a natural progression of a lot of the same stuff I was doing. It's kind of the new way for companies to focus on their core business values while not having to invest or worry about some of the other things that would generally cost quite a bit of money, be a distraction, and take time. To me, the whole cloud revolution or evolution came from complacency in IT. IT in the nineties and early 2000s, just wasn't really flexible, and didn't really focus themselves as partners to the business. It would cost a lot of money to swerve and change direction, and take weeks to deploy VMs in some extreme cases. So business units would have shadow IT departments dedicated to them. The cloud was really attractive to a lot of organizations that wanted to do a lot of things, one of them being Internet scale with the whole Web 2.0 thing coming out in the mid-2000s. And then you also had companies where IT wasn't really a strong business partner.

This is critical in today's ultra-competitive world where speed to market is more critical than quality in a lot of ways. Companies need to define the market in order to maintain relevance, and the cloud lets them do that in a much more flexible manner than traditional infrastructure. So we've forced companies to really take a look at the entire idea behind cloud because now the idea isn't necessarily when or if it's going to be used. It's when and how do you build enough structure and adoption around cloud to where you can still control what people are doing it, without having massive expenses and having it integrated correctly.

Mala: Describe a few things you wish you knew when you started your career that you know now, and you recommend to people that are new.

Dave: Don't be afraid to reach out to the community. Don't be afraid to be wrong. Be curious. And if you are wrong, don't take it to heart. Learn from your mistakes and experiences. Learn from other people. Talk to other people. Get to know other people that do the same things you do.

And even different things. The community is huge, and I wish I would have gotten involved in the community a lot sooner than I did.

Mala: What are the essential considerations one should have before migrating to a cloud computing platform?

Dave: There's a lot of them. It really comes down to what your overall cloud adoption strategy is, and what you're going to be doing with the application or the set of applications, and how you're really going to implement. One of the biggest mistakes that I see is customers simply moving workloads into traditional infrastructure as a service (IaaS) components into the cloud and expecting it to be cheaper.

It may not be, it really depends, there's a lot of investigative work you've really got to do to understand what your workload characteristics are and if they fit in the cloud and what portions of your application could take advantage of platform services in order to really gain a lot of the efficiencies in the cloud.

And the other part of this is the integration component. How do I integrate my identities? How do I integrate it in to my development workflow? What does all of that look like? So there's quite a bit to consider here. There's this whole concept around the five Rs of application, migration, and modernization: rehost, refactor, revise, rebuild, or replace. To me, considering the five Rs is the perfect place to start.

I think about it as a modernization and an application strategy. I'm going to take this application and move it somewhere because I have sufficient business reasons to disrupt the application. Or it's something new that I'm building that I want to be agile and fast and have a lot of scale.

It just really depends on what you're trying to get out of it.

Mala: How do you design for availability in a cloud-based environment?

Dave: As a consultant, I will never quote availability numbers higher than what the actual cloud platform will advertise, and I make it very clear that there are ways to design your application to have higher availability, but even in cloud that comes at a cost. It may not cost as much as if you were to buy all redundant hardware, redundant network links, redundant data centers, what have you, but there are ways to design for that in some cost-effective manners.

With Azure SQL Database, this can be as simple as a checkbox to enable multi-region high availability. You've got availability sets, fault domains and update domains, geo-replication, Azure Recovery Services, Azure Traffic Manager, Queues. There's any number of technologies for both IaaS and platform services to really do that, so it really comes down to understanding what the SLA requirements and the RTO/RPO requirements are for whatever you're rolling out and finding the appropriate technology to meet the need.

Mala: What can you say about security features and sensitive data protection? I think that's evolved quite a bit since the olden days.

Dave: I think that there's been significant investment in cloud security overall, and I think Microsoft has a really good story on the Microsoft Trust Center for what they've been doing with cloud security. I talk about the fifteen billion dollars that Microsoft has in its infrastructure, and unless you're Facebook or Google, you're not spending fifteen billion dollars in infrastructure. And then there's a lot of technology, like the web application firewall built into Azure or an application gateway. There's threat management and threat detection, and encryption built into Azure SQL Database. There's DDoS protection. There's also Azure Storage Service Encryption for your blob storage, and for VM disks.

I really think the security question at this point is more of a cultural problem than an actual legitimate concern in most cases, especially considering there's infrastructure deployments and policies, PCI, and HIPPA environments. A lot of platform services have gotten HIPPA compliant. There's also the federal government cloud that's available in AWS and Azure designed to meet federal requirements.

But that being said, it's still your responsibility to secure your application and to secure your infrastructure. Those don't go away.

The overall cloud platform may be secure, but you're still responsible for the last bits of it, making sure that you follow proper procedures and design decisions when build your applications. Make sure that you've locked down your environment, that you don't have ports open, that you've got strong passwords, all kinds of stuff. There's still a lot of reliance on security to the individual deploying and designing, but a lot of the other stuff that you would normally have to worry about really doesn't become a concern anymore.

Mala: How would you size a system right for normal and peak situations? How is that different from sizing a system with on premises?

Dave: There's a number of tools out there to help you with this. However, when migrating from on-premises, one of the biggest challenges with sizing is the limited number of machines that are available. Granted there's a lot of different types of configurations, but you've got a mostly unlimited amount of configuration options when you're deploying on-premises because you can back it with, for example, NVMe-backed storage with 160-gig interconnectivity on a UCS and all that other stuff. It really comes down to understanding what your baseline workload is, what your peak looks like, and then finding predictable patterns around the peak and what makes sense to deploy.

For example, virtual machine scale sets can be a good way to handle that because you can spin up more instances of that particular type of machine and start handling workload immediately and scale it down when you're done. One of the truths out there is that a lot of machines sit mostly idle, but I spent six years of my life working in eDiscovery. eDiscovery is all about designing for peak workloads because you never know when the peak is going to happen, so you need to design your systems with enough headroom to be able to handle increased capacity and demand. But I think the cloud makes that easier because you don't have to write a purchase rec to buy another virtual machine host when your VM environment is full, instead you spin up more machines and the nearly unlimited capacity around storage and machines it helps. But you really have to design your application, if it's a new application, smartly. You have to be able to break it up into services that can be independently scaled or just scaled on demand and only parts of the application scaled as necessary.

Mala: What would you say to a DBA who wants to monitor cloud-based systems? What are the metrics that he or she should be keeping track of?

Dave: It depends on the type of resource that they deploy. So if deploying Azure SQL Database, there are certain metrics that you're not going to have exposed—some of the more deeper operating metrics, but there's a different set of metrics that you'll have to look for. Where you are with overall DTU utilization, your space utilization relative to how much space you've purchased, and locking, blocking, maintaining indexes—those kinds of things don't go away. They just change in some of the ways that you deal with it.

And traditional IaaS infrastructure is very similar to managing what you're managing today. You look at a lot of the same metrics. It really is service dependent. However, there's a number of utilities and a number of graphs and data points that are built into, let's say, the Azure portal, for example, that would give you a pretty good overview of what you're trying to monitor, all the way down to the functional equivalent of Query Store.

And then with the cloud, you also have some self-tuning capabilities built into technologies, like Azure SQL Database.

Mala: When does it make sense to stop using a cloud-managed service and begin managing your own environment? Is there a tipping point?

Dave: I think it comes down to cost, and the ability to design for the cloud. because ultimately, with cloud, large monolithic applications are rarely successful because you have an upper-end limit to how much throughput, the number of IOPS, and the size of a single instance of a database can be deployed. You don't really have that problem with on-prem because you can always buy bigger hardware, including superdomes.

But the point is if you can't redesign your application to take advantage of the cloud's scale, then sometimes it does make sense to bring it back in. But realistically, I don't see that happening as often as a lot of reports might make you think. I'm pretty sure it does happen, but I think that customers are finding better ways to utilize cloud services without having to deploy things on-prem.

Mala: I've seen a few reports, and I thought there was a lot of that happening.

Dave: I've seen reports as high as forty percent. I think when you start looking at the details, you find out that it was due to the lack of strategy and lack of cohesion to begin with. I think it comes down to a lot of foundational problems.

Mala: What kind of diverse tools have you used, and which ones do you prefer?

Dave: Visual Studio Team, VSTS—very well known, very good. I've used Ansible, Chef, and Puppet—very good for their purposes. Terraform is also very popular. I also do quite a bit with ARM templates, PowerShell, and Azure CLI. Jenkins and some of the Redgate utilities are pretty great also. For monitoring, I like Stackify, Datadog, and a few others

Mala: What are some industry trends that you're really excited about and watching for?

Dave: For operating systems to become irrelevant. I think this is the ultimate goal with containers. I think that ultimately the answer is overall containerization of workloads and applications, and move the focus from the server and operating system to the service that it provides. I'm very excited about the containerization of workloads to be able to deploy the exact same image across multiple environments and multiple releases without any of the environment or configuration skew that you could run into very easily. It's a predictable way to have the same bits deployed in multiple places, predictably and reliably. And if you have a problem with the release, it doesn't matter – just pull the last image that worked and deploy that. It can be just that easy. You decouple the data from the compute, and the next thing you know, you have an instant upgrade cycle with much less effort than on traditional installations of SQL Server.

Mala: A lot of people I talk to said they were very excited about the containerization thing.

Dave: Yeah, to me that's the utopia we were promised in the late nineties and early 2000s when virtualization started really becoming a thing. All the way into the early to mid-2000s, that was the utopia we were promised for app deployments and app virtualization, and other things, and it's finally being delivered through technologies like Docker, Kubernetes, and Mesosphere.

Mala: What are the conflicts you experience when you interact with people? Other technologists, like SAN, developers, and network folks. And how do you normally resolve them?

Dave: I'm going to take a step back and say that in my role as a consultant, I've been the most successful by being able to talk to those resources and bring teams together.

My background is a little different from some other data professionals. I grew up in infrastructure. I've done a lot with Cisco, Juniper, and other network technologies. I've managed storage arrays, VMWare clusters, Web farms, and work regularly in Linux. I do have an application development background, enough to where I can speak some of the same languages.

I've been able to be the person to bridge the communication gaps between the database teams and the other teams. For example, I had one very large multi-billion-dollar customer based out of St. Louis where the DBA team was constantly angry because they weren't getting the performance that they needed out of their LUNs. I talk to the SAN admins and it turned out there was never a discussion around performance. It was always a discussion around capacity, and they were just giving what they were actually asked for, in a very literal sense in how they understood it.

They literally provided what was asked for because there was never a meeting or a team discussion to talk about what the general requirements were, and then having the other teams figure out as the SMEs the best ways to deliver that. In a couple of days, I was able to really work with the storage team, the network team, the server team, the operating system teams, the virtualization team, and the database team, so we're talking about working with about five or six different teams, to agree on when we request a SQL Server, these are the things that we're going to ask for. These are the questions the storage admins need to know. These are the questions the network team needs to know. These are the questions everybody else need to know. These are the group policies that need to be applied for performing volume maintenance tasks, lock pages and memory. We turned that into a request template in their ITSM utility. Now there's a template for them to request machines with certain fields that are required and you have to choose, and that turned into a big win because now the storage teams aren't getting mad, because the DBA teams aren't getting mad at them. Finally, we got people that worked just down the hall from each other to talk to each other for the first time in a number of years.

There can be issues when you talk to those individual team members if they don't have the right context, or if they've already had a lot of bad experiences with data teams—you know, not knowing how to ask for the things that they need, and them not understanding what the needs and the requirements are. So when you're able to bridge that gap, good things happen.

Mala: That's a great story.

Dave: Yeah, it's like nobody wakes up at the beginning of the day saying, "I want to suck at my job and do all the wrong things." Right? People don't generally do that. People legitimately want to do the right thing. People legitimately know that we're all here to accomplish the business tasks and business goals. It really is just about getting people on the same page and working together.

Mala: What are your favorite books, blogs, or other means of learning? How do you keep up?

Dave: I read a lot. I read a ton of blogs. I read a lot of Twitter posts. And I play a lot. I use my Azure credits to the fullest every month.

I have a home lab that I play in, and I run beta versions of operating systems because of some cool feature that I saw that I want to try out. So I do a whole lot of hands-on play.

But sometimes the stuff that I'm working on has a direct relation to something that could help a customer, so it works out.

Mala: What are your recommended ways of stress management that help a healthy work/life balance?

Dave: Okay, so this has been an interesting hot topic, especially considering Jeff Bezos's recent "people shouldn't seek work/life balance because that's too difficult." To me, I think everybody's different. I think everybody deals with things differently. Personally, I like to bury myself in technology and do things, but you also have to take a step back and say, "When I'm having lunch, I am in a different location, and I'm not reading my phone or checking email." Or after a certain time of the day or on weekends, you've got to make that time for yourself, and your family and your friends, and really be strict about it. Draw the lines. Sometimes things are going to go over, that's just the nature of the beast working in IT, but realistically, there's a certain level that can be controlled as well.

It shouldn't be about working to work. It should be working because it helps you be able to enjoy time with your friends and family and other things.

Mala: Describe your style of interviewing a data professional. What do you look for in somebody you're looking to hire? Aside from the technical questions.

Dave: To me, technical questions aren't as interesting because I don't think that they necessarily tell you much about a person.

I think that anybody can read any number of blogs or interview questions or interview looks and do relatively well there. I like to walk people through scenarios. I like to understand as much as I can about what makes the person tick and understand how their mind works, how they problem solve, and how they troubleshoot. You don't expect everybody to know everything at all times, and when you ask people silly little questions that are just nitpicky things, I think that says more about you than the person that might have answered it wrong.

To me, it's more about understanding the person and understanding how their mind works when it comes to problem-solving and coming up with creative solutions. Really listen to the person being interviewed. Really pick up on when their voice rises a little bit, and understand the parts that excite them and the parts that don't. Really weigh that to what you know the job position is going to entail and try to figure out if you think they're going to be a good fit culturally, personally, and professionally.

If I'm hiring for a particular immediate need, then skills are absolutely important. If I'm hiring more for a generalist, I think the fit of the individual is more important because any gaps in skill can be filled as long as the aptitude is there and their willing to learn.

Mala: What are your contributions to the community, and why do recommend people be involved with the community? I think we started with this a little bit, so can you say more on that?

Dave: I like helping people. I like sharing knowledge. I like teaching. I like doing things like that. I try to give back through, more recently, blog posts, but presentations at SQL user groups, at SQLSaturdays, and just helping people out with problems. I've got friends that call on me if they need an additional pair of ears to listen through their problem, and vice-versa. That's a form of giving back too because you're helping someone else be successful.

And to me that's amazing. I mean, being able to help someone else out is more important than me helping myself and learning on my own. I get excited about things, so I want to share with people the new cool thing I just learned.

Mala: Do you have anything to say with regards to documentation, how important documentation should be to a data professional?

Dave: I think documentation is absolutely critical. I think it's the thing that people like doing the least, but it's among the most beneficial things you can do for yourself, especially in situations where you want to take a vacation and you don't want to be called.

Key Takeaways

- Don't be afraid to reach out to the community. Don't be afraid to try new things. Be curious. Don't be afraid to be wrong.

- As a consultant, I will never quote availability numbers higher than what the actual cloud platform will advertise, and I make it very clear that there are ways to design your application to have higher availability, but even in cloud that comes at a cost.

- The overall cloud platform may be secure, but you're still responsible for the last bits of it around making sure that you follow proper procedures and design decisions when you build applications. Make sure that you've locked down your environment, that you don't have ports open, and that you've got strong passwords.

- You really have to design your application. If it's a new application, smartly. You have to be able to break it up into services that can be independently scaled, or scaled on demand and only parts of the application scaled as necessary.

- Large monolithic applications in the cloud are rarely successful because you have an upper-end limit to how much throughput, how many IOPS, and the size of what a particular single instance of a database can be deployed.

Favorite blogs: Microsoft's Azure Blog, High Scalability Blog, BuildAzure Blog, Docker's Blog

Favorite Websites: Azure Documentation, Microsoft Trust Center, Channel9 Azure Friday

Favorite Training: Hands-on Labs, Pluralsight, Azure YouTube Channel, Cloud Ranger YouTube Channel

Matt Gordon

Data Platform Solution Architect, DMI

Matt Gordon *has worked with SQL Server since 2000 and has worked with versions up to and including SQL Server 2017. He is the leader of the Lexington, KY, chapter of PASS, a frequent SQL Saturday speaker, a user group speaker and a speaker at PASS Summit 2017. He's also been named a 2018 Friend of Redgate and IDERA SQL Superstar. His original data professional role was as a database developer, but that quickly evolved into query-tuning work, which further evolved into a full-fledged DBA in the healthcare realm. He has supported critical systems utilizing SQL Server across multiple data centers and managed dozens of 24/7/365 SQL Server implementations. He currently utilizes those years of real-world experience as a national Microsoft data platform consultant helping clients design deployment solutions that meet their ever-changing business needs. He blogs at sqlatspeed.com and can be reached on Twitter at @sqlatspeed.*

Mala Mahadevan: Describe your journey into the data profession.

Matt Gordon: I started out with an accidental DBA gig. I think a lot of people have. I was working at a company, basically hired as general IT, and I think they had purchased an accounting package that ran what would have had to have been SQL Server 2000 at that time. The main IT guy really didn't do anything

© Malathi Mahadevan 2018
M. Mahadevan, *Data Professionals at Work*,
https://doi.org/10.1007/978-1-4842-3967-4_23

with databases, and my recollection is that the vendor support was quite lacking. Unsurprisingly, we started to have some issues with the software. They knew that I was straight out of school, and they asked if I took a course on databases and if I could figure out what was going on. And so that was my first effort on the operational side of being a DBA. I wouldn't call it database work per se, as it was just one server, the one app, but it was something.

That kind of got things pointed down this road. The next job I took, I was a support analyst for an asset management software company. I was doing pretty well with that. Our software at the time, you could run it on Oracle. You could run it on SQL Server. And the reporting platform allowed customers to write their own queries.

As you might imagine, not all of them were very good at it, and so we had a reference customer list where the support people were dedicated to a customer or two and giving them direct support. They wanted to create a group of us that were dedicated to supporting customers with this ad hoc querying ability that the reporting platform had. If they wrote something that was slow or didn't come out the way they wanted, or whatever, we were supposed to assist them with that. At that point, I had nothing but academic knowledge of database stuff, especially performance tuning.

But my boss—I have him to thank for where I am now, to his credit said "I think you'd be good at this. Why don't you give it a shot?" That went well and that led to me being what they call a release engineer in QA, which was not only tuning some stuff but also keeping all of our build environments up across different operating systems and different database platforms. The next job I took was a SQL Server DBA position.

And that's more or less what I've been. Of course, as a consultant, you're not really a DBA or an anything. You're a data professional. But that is how I got here.

Mala: Most people are actually DBAs, and then get into doing to do other things as they go along.

Matt: Yeah, when I came out of school, I thought I was going to write code. I had done some IT co-op things to pay my way through college, so my work experience was all general IT. And that ended up being where I started. That turned into this. I'm really, really happy it did.

Mala: Describe a few things you wish you knew when you started your career and you now recommend to people who are new at it.

Matt: Oh man, there's a lot I wish I knew. I wish I knew how dynamic a career in technology was going to be. Obviously, I was a computer information systems major, which at Clemson was basically computer science, but we took business classes instead of the lab sciences, so they called it something else. But I thought, "I'm learning to write code in C++ and that's what I'm always going to do. That's what I'm going to always need to know." I had no idea, and if I had, I definitely would have approached the academic side of college a little differently. I probably would've even scheduled different classes.

I'm glad I had the business background, that certainly seems better than having taken a bunch of chemistry classes, but I definitely would've taken a different approach to school knowing that it was unlikely I was going to walk out and be doing exactly what I thought I'd be doing.

That's the one main thing. There's other stuff, but that's really the one main thing because that would've changed some of what I did in those five years. So my recommendation is to be prepared for constant change. That's only accelerated when I was a DBA at a company supporting software that I write, that was fairly static. We were trying to stay roughly current with SQL Server, and of course, we would make changes here or there, but it was pretty static. You wanted to make sure you kept the lights on, and that you weren't running out of storage or servers doing what you needed to do. But that was about it.

When I got into consulting, obviously that all changed, because I'm working on a lot of different stuff all the time. Maybe it is because I am new to consulting, or maybe consulting just opened my eyes to it a bit more. The pace of change seems like it's accelerated over what it was even three years ago.

I guess some advice to any newcomer would be don't pigeonhole yourself. Even if you're just starting out, don't say, I only do BI stuff, or I only write reports, or I'm only a DBA. Those lines are blurring all the time, and if you say that, not only are you going to keep yourself from the knowledge that's going to be useful down the road, you may limit your own career options as well.

Mala: What was analytics and data visualization like even five years ago? What has changed, and why is it such an in thing now?

Matt: That's an interesting question. All in all, it was a lot less advanced. You couldn't do things at the scale you can now. Obviously, "the cloud" has helped that quite a bit. You have the ability to spin up some really powerful environments without having to buy anything up front. You just go on, click a few buttons, and then you've got a whole bunch of horsepower. You don't need to have a bunch of meetings planning it and have racks shipped in and take weeks to hook those up. It's just we have it now.

So that's obviously changed it quite a bit. I feel like the data visualization stuff has changed in parallel. And maybe for the same reasons, but I also feel those tools have become not only more powerful but much more user-friendly. Of course, most of what I work on is Microsoft. I do know a little bit of Tableau, but I've been around more people doing Power BI work. It's easy enough to use, and Tableau's pretty similar in this way, where you see people who really don't have a technical background at all but they do have good knowledge of the data that they're looking at. For the most part, the tools weren't mature enough. A user couldn't just sit down and make powerful visualizations in just a few minutes. Now, however, they can do that. The tools are designed more intelligently, they're easier to use, and they have much more power behind them to make some of this stuff work.

I guess that bleeds into the second part a bit. What's changed is the scale and the buzzwords used. So there's a lot of people talking about doing analytics, and they're not. They're doing stuff with data that's the same they were doing three years ago.

Now it's cooler because "we're doing operational analytics," but it's the same reporting you did thirty-six months ago. You just call it something else. I think the main thing is that a lot of this stuff is becoming easier and easier to use, and because of that, it's becoming more powerful to a company.

Some of this stuff I've done with sentiment analysis and Azure Logic Apps. I started doing it just as a silly thing for a soccer podcast. I sat down, thought through it, and thought of making a presentation out of it. What that's turned into is really powerful, and when you combine it with some other data, it's interesting. Like I mentioned, I've co-founded a sports analytics company.

And folding that kind of stuff in where we can, depending on the level of interest of the franchises or the clubs, we can overlay player analytics that they already have with real-time reaction from social media.

Then you can do this with a marketing campaign: here's who you're reaching, here's where they are, etc. All that information is a lot easier to get than it was a few years ago. We're very new, and we've just worked with a couple of clubs, I can't even tell you who they are, but we have all this information and we're constantly building on it. A lot of these clubs and franchises collect a lot of data but didn't know there were these other pieces, and it's massively changing all the time. They promote their matches and games the way they promote their clubs and souvenirs, but we can help them engage much more directly with their fans. If we had thought of doing this three or four years ago, at a minimum, it would have been a lot harder. Realistically, it wouldn't have been possible the way that we're doing it now.

Mala: What can a technical data person present to management? And as far as BI of analytics goes, how do you work past the buzzwords and present realistic ideas?

Matt: That's hard because it's often management who the buzzwords work on. They just see an advertisement and think, "I want big data. How can we have that?" They may not even know what it is. I'm not sure that everybody knows really what it is.

Because stuff is so pushbutton now, or at least it can be, I think people should be a lot less worried about presenting things to management as a proof of concept [POC] because as long as you have a decent understanding of the cost, the barrier to entry is low. There's the danger in Azure, or AWS, or whatever. If you spin up something you don't understand well, it can cost a bunch of money. So you need to be cognizant of that and be careful diving into some of this stuff.

But the barrier to entry is a lot lower, so you don't have to go to your boss and say, "I'd like to see what this analytics project can do for us, but to do this project, I need X servers, and I need Windows licenses and SQL Server licenses, and who knows what else. And oh, by the way, it's going to cost $100,000 and I still don't know what we're going to get out of it." If you do that, you're not going to get very far.

But if you say, "I need $5,000 to spend. I'm going to have this spun up, and I'll have something back to you in a couple of weeks," that can be pretty compelling if you're in an environment where management is receptive to things like that.

So if you have an idea, don't be afraid to put something together to take to management and leverage the ability that these cloud platforms have to lower costs greatly. And I think as long as you present something tangible, that should cut through all the buzzword stuff.

Now depending on the manager, you may want to buzzword it up so it sounds like something they're into. Some managers love that stuff and some don't.

Mala: What has been your experience with agile and business intelligence?

Matt: That is an interesting story. At one company, I was a technical and people manager for two groups—a database development group and a data warehouse development group.

Our main product was basically a system that processed WIC [The Special Supplemental Nutrition Program for Women, Infants & Children] benefit transactions. It's a government program for moms who have kids under five, and are breastfeeding or are pregnant. They go to the county health clinics as part of the check-up process, and they can get WIC vouchers that allow them, in theory, to buy healthy food for them and for the kids. It's not food stamps. It's actually a prescription. They meet with a nutritionist and walk out with what looks like a check that says one pound of potatoes, one pound of cheese, a dozen eggs, and so on.

Our software basically took that online, and now it's running in a bunch of states. We do analysis on this data and find transaction patterns that might indicate fraud. Things that would save the state agencies money. So you can walk in a door and say not only do we have the system to process all of these transactions, but we have this environment that goes with it where you can not only see in great detail what's going on but also help keep your program from being defrauded and costing more money than you expect.

So that's the whole setup for why we had a second spin-off team. There were great differences in opinion on how we could handle agile processes. Every other development team in the company was fairly similar. They were all developing data-driven web apps. There was a database developer or two embedded in every team, and then you tried to make those folks cross-functional so they could write some basic .NET code to help the app folks.

These were, more or less, your classic cross-functional scrum teams. On the BI side, I had pretty strong feelings that we could have situated that group in a similar fashion and some of those above me disagreed. There was a lean toward saying "we're agile" but it was a cover for "we don't have any defined processes."

It is doable. I've been in places where you have BI teams that are high functioning agile teams. It is tough though. It is a different method of development in a lot of ways, and especially if you're out of a company where you're taking people out of more traditional app development, and you're saying let's design a data warehouse or let's model this data. You definitely would break the stories and tasks down in a different way. And, unfortunately, some people are just not going to do it.

Mala: What has been your experience with cloud adoption?

Matt: I do a lot of talks on high availability, and a lot of the salespeople from Microsoft, Amazon, and all those big companies are really good at basically saying, "The cloud is magic. It never goes down. It never breaks. You never lose data."

And that's not exactly true. You still need those same thought processes, the same ones when you're designing something on-premises. It requires the same thinking. You just have different tools at your disposal. My experience is that sometimes places dove head first into the cloud stuff and bought every line that the salespeople had. Then they were really disappointed when something broke and they would go back to the salesperson and say, "You said this would never break!"

I've also been at companies where their particular workload or whatever they were trying to do fit very well, almost perfectly, into something cloud-based. I see that with BI quite a bit because you've got things like Azure SQL Data Warehouse. We know we need a warehouse to store this data but it entirely variable how much data we'll have. We won't know for a year, but we have to start in three months. In fact, during my cloud presentation at PASS Summit last year I ran into this. There were people interested in my talk who came up afterward and said they worked for a finance company in New York and are not allowed to use any of it because it's not secure. I pointed out the DoD [US Department of Defense] uses Azure. There's a lot more security there than your average on-premises data center.

That message has not gotten out very well and I'm not sure why. So I guess my overall experience would be kind of what I started out saying. It's been marketed too well and also not well enough around security. That said, as a consultant I almost don't go to a customer without some sort of cloud adoption being part of the engagement. That's a big difference from even a year and a half ago.

Mala: What are the challenges with analytics today in technology and the community?

Matt: I think the technological challenge is there are so many different platforms, and there are so many different tools that it's easy. But you end up stagnant because there are fifty different ways to do it. You may not know which one is best, and so you're just not going to do any of it. Community-wise, in the Microsoft community, I think we're spoiled because the #SQLfamily stuff seems to have naturally and easily grown to encompass all the other stuff that SQL Server can do including all the stuff we can do in Azure. It feels like it's been fairly natural. SQL Saturday`, PASS Summit, and other Microsoft data platform events are not just a bunch of DBAs and database geeks. We've got analytics folks, analytics tracks, etc. It's been a natural outgrowth.

I think maybe for other platforms, that maybe that community is not as well organized. I have lesser experience on the Oracle side, but people I've worked with that have more experience with some of the other vendors and other platforms, do kind of look at the Microsoft data platform community pretty jealously because they think we've got it pretty good.

As a PASS local group organizer putting together our meetings, I can go to my group and ask them if they're alright with having a machine learning topic, an analytics topic, traditional DBA topics, and BI talks, and everybody's cool with it. They just think this is all really interesting and this is all part of the greater community of what we are and everybody regardless of job title, they still show up, because they're interested in it and know they might need to know it.

The main challenge is that there are so many different ways to do this. You see it on Twitter, but even a lot of this stuff can evolve into arguments between nerds – "this is the way I do it, your way's stupid, well your way's stupid" and so on. If they both work it's just really what you're comfortable with. I've definitely gone to some customers especially here recently where they just don't do anything and they say they're going to wait for things to shake out. I don't know if they ever will, or if there ever will be a less chaotic way to accomplish these things. When they put R in SQL, they're going to keep adding. I don't ever see a time where they're going to be say that now you can only use Python for ML stuff in there. We took R out.

People sit on the sidelines saying they're going to wait for some of these things to wane. I'm not sure they're ever going to wane. I think they're going to keep getting better. There will be different and better methods. I just think there will continue being multiple ways to provide these data solutions.

Mala: Some people think storytelling means giving stories about data, and data should just present facts. What's your take on that?

Matt: Well, it really depends on who you're talking to. What I found interesting about being a consultant is that you get to interact a lot at the different levels, you're with the tech team a lot but you're also likely talking to managers or even executives as well and, depending who you're talking to, they're going to want to see it presented in a different way. Executives, especially, that were technical before they rose to that level, are generally bare bones in terms of data. They just want it in black and white. We asked a series of questions. Tell us the answers.

With other parts of companies, there is more of a story you're telling. Sometimes it depends on if you're trying to persuade an organization to make a change in something. Some of the stuff I was talking about with getting a professional soccer club to change the way that they market certain things, you have to bring that along like a story.

Now you're going to underpin it with where we tested this change, you guys sold fifty percent more T-shirts than you did across that same campaign last year. You're going with hard facts, but you're still going to bring them along with a story. Why they should want to sell more of those shirts or have people out for this kind of promotion or whatever. Definitely, there are some types of people that you're talking to where the data is just black and white facts and they're always there but you do have to kind of colorize them a bit.

Mala: What are some of the data quality issues in dealing with data for analytics? What can be done to minimize them?

Matt: Again pointing to one of these professional clubs we're dealing with, they're reliant on their league to provide the data on souvenir sales and that solution appears to have been designed by somebody without a good understanding of how that data would be used(to put it nicely). So, you're matching orders and quantities against long text description of items that may only differ in the last characters or two, instead of order IDs, order detail ID, common things.

That complicates greatly the club's job in coming up with reports on this, because they have a data group at the club but they're not being handed data that's all that easy to use and whatever way you're going to do to match that stuff up, is going to be error-prone, because you're kind of fuzzy matching some things, basically.

So that's one challenge. Whatever's gathering the data wasn't designed with the data's end use in mind. Another thing is that if it's not well validated, you're going to end up with a bunch of junk anyhow.

You're likely leaving it to the end to somebody who maybe is only on the team to do reporting or data visualization or something like that. They're probably not data qualified but the job is left to them, they know enough to say well this looks like it might be garbage and I'm going to try to clean it up.

They may or may not be best suited for that. But yeah, a lot of it is, I think the point is those solutions are not always designed by people who think about how any of that data is going to be used. They're just trying to get it in, and it's spewing garbage to the data folks.

I've seen that more than I ever thought I would. And it's not even a technical issue. It's just that there wasn't communication group to group. "This looks nice so I'm going to do it this way," but it's not at all serving the needs that it's supposed to.

Mala: What are some of your favorite tools and techniques?

Matt: I'm still pretty taken with what Azure Logic Apps can do. I know that it doesn't fit into a DBA mold. It doesn't really fit in a BI mold, but some of the things you can do in there are very cool. The social media sentiment has been interesting to me, and it has opened doors for me to give talks in some cool places as well, but when you take everything that can and start combining it with Azure Bot Service and other things, that's so fascinating to me.

I've also started having more engagements around combining Power BI with Azure SQL DW, that's pretty cool as well. I guess I would describe those as my current favorites. They may not be for everybody, but from a technical nerdy standpoint, there is cool stuff going on here.

Power BI gets better by the month, and I know your hardcore Tableau folks would say that it hasn't caught up to Tableau. Maybe it hasn't. I think both tools are really good.

Mala: That's what all the folks used to say of Oracle and SQL Server.

Matt: The Azure SQL DW and Power BI combination is for the right customer, but it can change their world. And it's really fun to work on when somebody else's paying for it.

Mala: What are some of the new technologies in the big data BI world that you're keeping up with and excited about?

Matt: Not to pigeonhole it by vendor, but Microsoft is most of what I do. I am really excited to see the Azure SQL DW evolution on the BI side of the fence. On the DBA-ish side, I'm excited to see the opportunities that Azure SQL Database, and then the Azure SQL Managed Instances offer people on the more traditional relational side of the fence. Combining a managed instance using PolyBase to hook it to some of the semi-structured data store that's putting a ton of power in people's hands, which was very difficult to near impossible to put together before SQL Server 2016 came out. And now we've come so far from that that you're giving organizations a lot of power combining all this stuff, and I'm really interested to see where it goes.

Mala: What's the role of documentation in being a good BI/analytics person?

Matt: A necessary evil. I went to school with a guy who used to say, "Comments are for the weak-minded." It was funny, but I disagree. A lot of tools change, so I think documenting what your solution is and why you did what you did is even more important because the way things go these days. It is possible that you may create a solution sitting here in the summer of 2018, and you're going to use any of the tools that I just mentioned.

And in the summer of 2019, maybe you're going to reimagine that with something else. In fact, it's more and more likely that that will happen. I've worked with customers that were all in for Azure SQL DW a year ago and are now saying, "Azure SQL DW is really cool, and we like it, but let's explore Snowflake now."

They want to revisit what we did with it. I think that's going to become more common, and it makes the documentation that much more important. Because if you can go back and say, "Here is our thought process. Here are the documents we put together explaining the workflows, etc." And maybe explain some of the reasoning behind why it worked the way it did. It's just that much easier to redo it with a new set of tools.

If you don't have that, most of the time, somebody's going to end up explaining to the boss why it took twice as long as they thought because the team forgot how they did everything. Generally, bosses aren't too impressed with that answer. They're going to ask you why you didn't write it down!

I don't know a lot of technical folks who enjoy writing documentation, but you've got to do it. Even if you don't care about anybody that comes after you, just do it for yourself. If you do something and step away from it, you could come back six months later and say, "I don't know what any of this is. Why did I do this?"

It's useful to have stuff written down to explain all of that.

Mala: What are your favorite books, blogs, and objects of learning? How do you keep up?

Matt: I love Twitter, and I hate it. I've had it for nine years. My little Twitter anniversary thing came up recently, but I rarely used it. Usually, I would follow a sporting event that I couldn't see on TV. They would tweet out what was going on, and that's all I ever used it for.

When I got into consulting, my boss at the time said, "Now that you're a consultant, you need to be on Twitter. You need to be following the right kind of people, and you need to be active yourself."

I didn't want to do that, but he was one hundred percent right. Sitting here and thinking about what blogs I visit every day or every other day, it's a list that constantly changes. As I've been more involved in the community, I start to figure out which people follow certain things. The blogs that I might be into at any time are likely a reflection of a customer engagement or something that's caught my attention. I can't sit here and honestly say here are the blogs I read every day because it's dependent on what I see that interests me with the people that I follow on Twitter.

There are many little subcategories of interest. When I was a pure DBA, I had a roster of blogs that I read always. Now I don't because it's just so fluid.

Mala: How do you handle stress?

Matt: Bourbon is handy. As a Kentucky person, we love those tax dollars! I mean, finding a way to get away is good. I've been fortunate enough where I've had racing the last several years, and that's really a consuming thing when you're there at the track. I'm not driving the race car thinking "Geez, why did I write that one piece of code that way?" I can't. If you're doing that, you're going to crash. I've only done that a couple of times, and it's no fun.

Occasional vacations, of course, are always great. Sometimes life gets in the way of that, but some other hobby that consumes your mental space is key. That's been racing for me or going to watch live music. I'm fortunate enough that I've got a group of friends where we all like similar music, so that's an occasional option as well. Sometimes just a night out at a venue seeing a couple of great bands, that kind of takes your head away.

But something that's distracting enough where you're not thinking about work is critical. That's going to be different for everybody, but racing and music is what does it for me.

Mala: Describe your style of interviewing a data person. What do you normally look for in somebody you're looking to hire?

Matt: Boy, that's an interesting one. I've evolved a lot on this from when I started doing it. We don't always get the right people. Technical knowledge is important, and it can be challenging sometimes in an interview to get that out. Where my interviewing goes, we would always start with a few knockout questions so you knew if it was worth the time. Ask two or three questions that a person applying for that position should know, and if they know those, we're going to move forward. What we found useful—especially since I was interviewing to place people on agile teams where communication was really important—is to come up with technical scenarios for them and ask them to go to the white board and show us what they would with it. We would produce some code, and tell them, "It's supposed to do this, but it's broken. Tell us how to fix it." Or give them a scenario that happened and ask them to go up to the board and talk us through how they'd go through it.

And then we would play the role of the team members saying things like, "Oh, that's a great idea." "Oh, that's a bad idea." "What about doing it this way?" And kind of get a sense of that person more in a work environment. I interviewed some people when I first started managing teams, and it was just, "Here's thirty things we think you should know." Sometimes they got them all correct, but they were completely non-functional within the team. They didn't communicate very well, and they weren't always as quick on their feet as they needed to be to work in an agile team. Based on that, we started to mold the interviews to simulate a workday as closely as we could.

I know it's hard to do, but we did it because the trivia contest method was getting us smart people, but it wasn't getting us the people we needed for the teams that we were putting them on.

Mala: What are your contributions to the community, and why do you recommend people be involved with the community?

Matt: I contribute what I can by speaking and running the PASS local group here in Lexington. Why do I recommend it? It can change your professional life. I went to SQLSaturday New York in 2015 and knew a couple of people there. They introduced me to a couple of other folks that have changed the course of my career. None of that happens without being involved in PASS. It's always a little bit more special to me to get picked in New York because I love the city but also because that particular SQLSaturday and the people involved in it changed my life.

Whatever role you want to play, it is there—quiet volunteer, speaker, organizer, etc. I certainly never fancied myself any sort of a public speaker at all. There are many other ways to get involved that people are comfortable with that make a difference. I was mortified of public speaking for thirty-five years. My life would be very different from what it is now without involving myself in the community.

Mala: Can you narrate a funny or interesting story?

Matt: This isn't funny, but it might be interesting. Twelve years ago, during my first full season of Formula First racing, I was third in our season championship going into the final two races of the year. You always put the car through tech inspection before each race weekend so the stewards can ensure the car conforms to the rules and that the safety measures are up to par. Going into this final weekend, my car had passed tech inspection everywhere else with no issues. On that particular weekend, we pushed the car into tech. I sat in the car. We checked weight, rain lights, etc. The inspector, a veteran steward, looked at me, the car, at me sitting in the car, the design of the roll bar, and made a face. To set the context, I have an open-wheel car, so my head is exposed. He said, "You're sitting too high up in that car. You need to change your seat padding." He said it'd be a one in a million thing, but you need to change your padding to bring the crown of your helmet down two or three inches. I think he was going to pass me through tech inspection regardless, but he said he'd feel better if I was sitting a little lower in the car.

We corrected it, as I had the shot to win the championship, actually.

Yeah, it didn't end up that way, and I finished third. Anyhow, we get to our last qualifying round for the season, and I see the one-lap-to-go sign at start/finish. I said to myself, "I've got one lap here, so I'm going to try something in the corners I hadn't tried yet and see how it goes." I'm either going to spin out or go faster and start closer to the front.

This particular track is really fast and flowing. I went into the first corner and tried something a little different. That meant I hit a bump that I hadn't hit any of the other times and it set me sideways. Now, the car never spun, but I did this long skid through the corner. As I was gathering it back up—so I'm almost facing forward at this point, all of a sudden, basically the back of the car explodes.

And I look up. A car that didn't avoid me the way it should have hit me. As I mentioned, we had open wheels, there were no fenders, so when the tires touch each other, you're usually going up in the air. A guy hit my right rear tire, and his left rear tire hit the top of my helmet on the way over.

It actually cracked the outside shell of my helmet. I was really sore, but I was fine. I'm sitting there a bit fuzzy and realizing that I got hit in the head, and then I thought back to that inspection. If that guy wasn't paying close attention…. I'm probably sitting four inches higher in this car, and what is just a really, really sore neck maybe could have been a lot worse.

I don't know that that's a funny story, but the fact that I'm here to talk about that wreck is interesting to me.

Mala: No, I think that's miraculous. I think that's a great story.

Matt: To kind of apply it back to work, there's nothing wrong with revisiting what's been done. Like I said, that guy looked at the form that goes with my car and saw I had passed tech inspection everywhere else, but he saw something that bugged him and he wanted it fixed. Professionally, that's good advice. You should never say, "We did it this way the last five times, and so it's probably fine." If he hadn't done that, maybe I would have broken my neck in that wreck. I don't know.

So that was always a bit of a wild story. And as a side note, I wanted to get the helmet back because it had a crack in it and a tire mark on it. I thought it looked cool. And unfortunately, in the mess of cleaning up the accident and going into the ambulance and getting checked out and all that, my helmet disappeared.

Somebody took it, and I don't know who because you're not allowed to use them after you get hit. But it was going to be a conversation piece here at the house, and it disappeared. Bummer, but at least I have the story.

Key Takeaways

- Don't pigeonhole yourself. Even if you're just starting out, don't say that you'll only do BI stuff, or only write reports, or you are only a DBA. Those lines are blurring all the time, and if you say that, not only are you going to keep yourself from the knowledge that's going to be useful down the road, you may limit your own career options as well.

- If you have an idea, don't be afraid to put something together to take to management and leverage the ability that these cloud platforms have to lower costs greatly. I think that as long as you present something tangible, that should cut through all the buzzword stuff.

- People sit on the sidelines saying that they're going to wait for some technologies to wane. I'm not sure they're ever going to wane. They're going to keep getting better.

Favorite tools: Azure SQL Data Warehouse, Power BI

Blogs to follow: SQLServerCentral.com, Brent Ozar's blog, sqlatspeed.com

Jimmy May

Storage Architect

Jimmy May has been providing solutions using SQL Server since the 20th century. He specializes in flipping the /faster bit. Jimmy is a Microsoft Certified Master (MCM). Formerly, he was a principal architect at Microsoft and a senior program manager for the SQL Server Customer Advisory Team (SQL CAT), where he managed the Customer Lab, which hosts the biggest, fastest, and most interesting SQL Server apps from all over the world. More recently, he was a SQL Server technologist for what was Fusion-io's Data Propulsion Lab (DPL). Jimmy has served as an enterprise architect at five organizations, including three Fortune 100 companies. He has spoken on four continents, including the largest data analytics conferences in North America, Europe, and Asia. At Microsoft, he was honored with a Gold Star award and a Challenge the Industry award. He is also a recipient of the Innovate the Industry award. He is a founder of IndyPASS and The Indiana Windows User Group (IWUG).

Jimmy recently moved from Microsoft's global headquarters in Redmond, Washington, to Park City, Utah. There he indulges his appetite for fitness. He's a fan of yoga and plyometrics, and he'll be happy to plank spontaneously with you—be prepared to stay horizontal for at least seven minutes. You'll find him at the local Orange Theory Fitness studio several times each week. He practices yoga, and he enjoys indoor cycling and mountain biking. In the spring, summer, and fall, you can find him hiking

© Malathi Mahadevan 2018
M. Mahadevan, *Data Professionals at Work*,
https://doi.org/10.1007/978-1-4842-3967-4_24

and biking the majestic mountains that surround him. However, his passion for the mountains is at its peak when they're covered with fresh powder and he's flying downhill with friends on a pair of fat skis. #PrayForSnow

Jimmy can be reached via email at jimmymay@outlook.com and on Twitter @AspiringGeek.

Mala Mahadevan: Describe your journey into the data profession.

Jimmy May: I did my undergrad in biochemistry. After my post-grad work I was doing some part-time work teaching CPR at a hospital in Indianapolis. It was all paper-driven. We had three computers in the department running a really cool new OS called Windows 3.1. The two PCs with Excel were always occupied by administrative people. I found a computer in the corner that didn't have Excel but had what looked to me a lot like a spreadsheet—something called Access. I thought I could adapt it to my purposes. It came with hard copy documentation—a book, which I took home each night to get up to speed on this concept called the database. The rest, as they say, is history.

I applied those skills to a job at an internet startup in California. They packed me up, and I drove all the way across the country to work on-site. Sadly, despite a lot of hard work from a lot of great folks, the business failed, so I packed myself up and moved all the way back.

Up to this time, I'd never even heard of the product known as SQL Server, but because of my Access experience, a headhunter brought me in to interview with an internet financial provider. I remember the first time sitting down and looking at the interface. What is this? Where's the UI?! I was hired on the spot, and in just a few weeks I was at the top of the help desk team, running it, training the team, optimizing existing queries, and writing new code to deploy to banks all across the country.

Things were going great until we had a catastrophic product release. I worked day-and-night for months to triage the disaster, earning a bonus that quarter equivalent to my salary, which they refused to pay.

I wasn't playing any games here. It wasn't like I took advantage of the system. I worked really hard, saved untold numbers of clients, and I truly saved their respective behinds. I did the hard work of CYA—cover your caboose—for them. Yet they wouldn't pay me.

Wow, indeed. It was unbelievable. I asked, "Are you sure about this? You're going to have a big hole to fill when I leave." "Sorry, we can't afford it." "How can you afford not to?" I quit. Not fair? Sure, it wasn't fair. But if life were fair, horses would ride half the time. I moved on.

Within two weeks I had a new job at a new startup paying almost double what I was making at that old job. It was a really cool gig. It turns out to be a company founded by the guy who invented voice mail—seriously, the inventor of voice mail. They explicitly told me I got my foot in the door there because of my enthusiasm, and I had SQL Server certification. They knew I didn't know everything, but they trusted that I'd figure it out. Of all the candidates who applied, those two things are what nailed the job for me. Certification has been a big career driver for me. I hope we get to that.

From there I went to another startup called ExactTarget, which is now a wholly owned subsidiary of Salesforce. I was employee number thirty-two. Why didn't I retire after they went IPO? Well, too bad I didn't have a better handle on negotiating stock options!

From there I was recruited by Microsoft Consulting Services [MCS] where for about four years I worked from Indianapolis parachuting in and out of cities in the United States and around the world. Somehow I found the time to research, validate, and evangelize a concept called disk partition alignment. That was the topic for my first ever presentation at PASS—to an overflow room of several hundred—no pressure!

As a consequence of that notoriety, I got a phone call from an organization called the SQL Server Customer Advisory Team—which we all know and love as SQL CAT—about which you know very well. Tom Davidson—inventor of wait stats—had sat in on my preso. He looped in his colleague Denny Lee and told him to get this guy—me—to write this up. Together, Denny and I wrote the white paper[1] that changed the world.

As you know, that's really what cemented my reputation in the community. Even now, a decade later, people come up to me. "You're the disk partition alignment guy."

My hero, Mike Ruthruff, is the one who actually reached out and invited me to join the SQL CAT. Even though I was nominally a storage SME, I had little hands-on hardware experience. But I learned! I upgraded our hardware and modernized our protocols. Under my leadership, the CAT Lab acquired dozens of new servers, three new SANs, implemented massive power upgrades, enabled parallel and remote engagements, and proved out Always On availability groups as well as columnstore indexes.

From SQL CAT I went to work as an architect with MSIT. The big takeaway there was introducing game-changing columnstore indexes to our warehouses and forklifting thousands of apps to Azure.

---

[1] http://technet.microsoft.com/en-us/library/dd758814.aspx

From there I went to a company called Fusion-io where I worked at the Data Propulsion Laboratory. In the last two years or so, Fusion-io was acquired by SanDisk, which was acquired by Western Digital.

I'm currently developing SQL Server content for MS Learn in collaboration with edX and Stanford University to be served up as a massive open online course [MOOC].

Mala: Can you explain why and what should a DBA really do to evaluate if he or she has the best storage they need?

Jimmy: A few years ago, I developed something called May's I/O Mantra:

$$X \text{ capacity (GB)}$$

$$\text{at } Y \text{ IOPs (transfers/sec)}$$

$$\text{at } <=10\text{ms latency for OLTP data files}$$

$$\text{at } Z \text{ throughput (MB/sec)}$$

$$\text{at } <=30\text{ms latency for DW data files}$$

$$\text{at 0ms-2ms latency for log files}$$

Jimmy: Let's talk about this. Classically—and sadly—people ask for and are assigned storage simply in the context of capacity. Yet performant storage cannot be defined simply in the context of capacity. Other relevant parameters include IOPs, throughput, and especially latency.

Asking simply for a terabyte of capacity is only the start. Storage is multifaceted. You wouldn't go to an auto dealer—ask for a red car, or a car with seven seats, or a convertible—and accept the first thing that was offered. In the context of storage, you ask for not only capacity but also performance in terms of IOPs if it's for an OLTP type of workload, or you ask in terms of bandwidth if it's for a data warehouse workload. And you absolutely *must* ask for that specification to be delivered at a given amount of latency. With regard to latency, the end user doesn't care how many IOPs you're getting, how much bandwidth of storage it has, how big the volume is. They care only how fast the app responds when they click the button or tap the icon.

Another critical issue is log file perf. You—and your storage team—need to know whether you're asking for a data volume or a log volume because the latency requirements are far more strict for logs. May's I/O Mantra capsulizes that much more succinctly.

Something that the Mantra doesn't specify but is also very important is resilience. How mission critical is this data not only in terms of performance but also availability? That opens up a separate set of considerations.

Lastly, validate your storage acquisitions! How accurate are the promises from the shiny-shoed commission-based sales geeks? Make sure their promises are testable. Get commitments for vendor assistance to remediate issues. Make sure your own team is on board with this. Build time into your deployment project for it. Validate end-to-end *before* flipping the production bit. Engineering discipline demands it.

Those are the things you need to consider.

Mala: Getting siloed into what we do is very common in our line of work. What do you recommend to ensure that DBAs have a more full stack perspective?

Jimmy: The answer I'm going to give you is going to differ from the conventional wisdom. This is often asked in terms of job security and relevance—especially in this day and age—consolidation, economies of scale, offshoring, cloud. Yet it's my view that we need to specialize—or at least have a niche of expertise. It's also important to be able to say no. For example, I make no apologies for never accepting a project having a statement of work including the term *replication*. Not that there's anything wrong with replication. I'm glad we have replication SMEs! My point is that I can't be an expert at everything, so I've chosen my areas of expertise with intention.

Jimmy: I made a decision to specialize specifically not only on SQL Server, but on SQL Server performance, and drilling down even more to SQL Server I/O perf. In fact, I drilled down even further by specializing in cutting edge I/O perf. When DMVs came out, I made it my business to get good at the perf- and storage-related ones. I focused on them. I/O remains the big bottleneck in SQL Server by an order of magnitude. My expertise in perf led me to getting that call from SQL CAT. When Flash was introduced, I acquired the expertise necessary to evangelize to the industry. I didn't go to work for Fusion-io because I wanted to work on Flash. Fusion-io courted me because I was an expert on Flash.

Mala: Specialization.

Jimmy: Yes, specialization *was* key to my success. In fact, in a former life a guy told me to be the best guy in the room at something. I took that to heart, and it's paid off for me very well. However, that having been said, it doesn't hurt to cover your bases! Look for what's going to be relevant. How can you as a data specialist provide value to your company? All but a small set of companies tend to be slow adopters. If you can provide value because of, for example, your data visualization, your analytics, or hey, there's this new data science thing, whatever—if these things intrigue you *and* you find ways to provide value to your employer...well, you can by intention define the job you love.

Do those in the context, I advise, of providing documentation of your learning. Formal Microsoft certifications, the certificates offered by Stanford/ edX/Microsoft—for example, the data sciences program. There are other certification programs as well. There's a data warehouse in the cloud specification or big data certification, for example. It's one thing to poke around with a technology. It's another to have done the work to earn the cert—that piece of paper is one demonstration of competence, as well as initiative. Full disclosure—I'm the lead on one of their upcoming courses. I'm a huge fan of certification. It's provided value to me throughout my career, not just getting my toe hold in that one startup.

There's a third thing I wanted to say with regard to your question. This is very important. Do not be afraid to lose your job! Do not live in fear! I've spoken to so many people who live that way. In fact, early in my career, I did just that— going to bed each night wondering if tomorrow might be my last day at work. There's a rich-and-robust tradition in IT of "fake it till you make it." As my friend MVP John Sterrett says, "If you're one chapter ahead of your customer [or employer], you're the expert." Another friend, MVP Mindy Curnutt, has a great presentation on imposter syndrome. Check it out!

It's likewise important to have the skills and the confidence to be willing to walk if things don't work out. Mala, how many people have we heard talk about the job they're stuck at, the job they hate? Well, there's a lot of great work out there for those of us who are willing, enthusiastic, talented, and qualified. You know that better than I! Go find it. Look around until your dream job falls into your lap. That's happened to me again and again. Or create it from your current role. It's not just luck. It's a key part of living an intentional life. Here's one of my favorite quotes—and it's a great motto to live by: "You are a ghost driving a meat-coated skeleton made from stardust riding a rock flying through space. Fear nothing."

Mala: What are some industry trends in storage with regards to databases that you're excited about?

Jimmy: There're a lot, but I'm going to talk about a few things, specifically Flash and S2D. Flash adoption is slower than I expected. Not only is Flash a game changer, but NVMe is more so. The newer NVMe products are just incredible. I will tell you specifically more about something in a minute. You remember the old days. A decade ago, ten or twenty millisecond latencies were acceptable for data, as I stated in my I/O mantra. Yet today we're getting microsecond latencies. A properly architected system—even one with high throughput I/O handling gigabytes per second—can be built to sustain microsecond latencies. It's amazing. I have an anecdote for later.

Besides the NVMe, there is S2D—storage spaces direct—from Microsoft. It is a high performance software-defined shared storage without the million dollar hardware and equally expensive licensing fees. Very cool stuff. Here's something I transcribed from a Microsoft site: "Storage Spaces Direct uses industry-standard servers with local-attached drives to create highly available, highly scalable software-defined storage at a fraction of the cost of traditional SAN or NAS arrays. Its converged or hyper-converged architecture radically simplifies procurement and deployment, while features such as caching, storage tiers, and erasure coding, together with the latest hardware innovations such as RDMA networking and NVMe drives, deliver unrivaled efficiency and performance."

S2D is compelling stuff—and it deserves to be a disruptor.

Lastly, this isn't under storage-specific, but high performance storage combined with high capacity is enabling a number of promising technologies. Some of the data science stuff, machine learning, artificial intelligence, those are more than just buzz words. Some really cool things are being abled by high capacity high performance storage.

I'm especially excited about AI. This notwithstanding Stephen Hawking, Sam Harris, and Elon Musk consider AI to be the greatest risk to humankind. There's an awesome TED talk from Sam.[2] I hope our silicon-based overlords are benevolent. We'll make great pets.

So there. Those are the four things I'm excited about.

Mala: As a storage expert, what has been your experience with cloud adoption?

Jimmy: Ah, I have a preamble. I was on SQL CAT when it was rebranded Azure CAT. This was a few years ago, so I'm comfortable sharing my "inside baseball" insight. Look, obviously I'm a huge fan of the team. We had a mandate from the Microsoft CXOs to go all-in for the cloud. But Azure just wasn't yet ready for prime time. Microsoft was slow to the game there. One of my colleagues—I won't say his name—was in charge at the time of the initiative. He had an office right around the corner from me in Redmond. We had a huge monitor in our main hallway. Microsoft was playing catch up—we've all seen it before. Every day we would walk by that big monitor with all these analytics. A big dashboard. So many of the metrics were red—some yellow and the occasional box with green, but mostly red, red, red all across the board. The guy was discouraged—we all were. There was lots of nervous laughter. It's pretty well known at this point that SQL CAT—Microsoft—jumped the shark in some aspects—losing focus on on-premises a little bit too early and diving into the cloud too quickly. We knew that, and in time

---

[2]www.ted.com/talks/sam_harris_can_we_build_ai_without_losing_control_over_it

even the folks in the C-suite acknowledged that on-premise isn't going away completely any time soon.

Anyway, it was a fitful start. A bit of a challenge for us. I want to be clear here. I am not bad-mouthing Microsoft by saying this. Just an anecdote that happened several years ago that we can laugh about today. Obviously Microsoft is doing *much* better now. They're setting records every quarter in terms of their cloud deployment, cloud volume, and cloud revenue. When I look at the Azure portal now compared to the old days—oh, Mala, it's so unbelievably rich and robust, and full-featured compared to what it used to be. New functionally is introduced all the time. It's a pretty cool story.

I transitioned to MSIT as an architect. I was part of the team that forklifted ten thousand apps to the cloud. Steve Ballmer said we were all in, and boy, he meant it. That mandate rippled down to us. We were forced to devise a means by which to get these apps to the cloud—or else, and we did it. I gotta tell you, it was a pretty exciting time.

Oh, by the way, the new data warehouse implementation from Azure is pretty cool. For the first time you can easily create elastic and out-of-the-box partitioning across not only tables but *across servers*—hash or round robin partitioning across servers out of the box. That's a pretty neat development. That's something that we struggled with for a long time. Implementation was a roll-your-own operation until this new Azure SQLDW deployment. Apparently it works as advertised—pretty neat stuff.

Well, those are my stories.

Mala: What are some other ways you recommend to improve the current habits and procurement patterns of upper management and other decision makers with regards to storage?

Jimmy: Oh, this is great—I love this question. First of all, be proactive but be patient. Change can take a year or two, or more. It is not realistic to talk about vast storage acquisitions in the context of overnight changes. You might get lucky in terms of timing—getting a whiff of early planning of a refresh cycle, or better yet, a green fields app—but don't count on it. An ideal time is keeping your ear to the ground for changes and parachute in when your org is architecting out of a disk IO subsystem. Be proactive. You need to be able to identify current and future scenarios, whether there's a bottleneck. Be patient, but able to talk at least in high level terms about what kind of remedies there might be. Develop an effective elevator pitch. Do so not just in terms of performance, but capacity, growth, resiliency, availability, etc. Also—this is critical—apply your highly developed soft skills here. You need to find out who's in charge, who cares about these things, who holds the purse strings. You may need to get in on or create a V-team to discuss your next gen storage. Perhaps get buy-in to incorporate your intentions into your annual goals.

You need to learn a little bit about the hardware acquisition cycle. Again, it's one of those things where you must be patient. It might take a year or two to get what you need. You need to document it. You have to sell this, not in just terms of dollars. For example, their proposed storage solution may cost less than another solution. Yet are the financial savings real? Are you really getting what you pay for? Are you buying a solution that actually fits your needs? Remember what I said earlier about validation! A lot of this stuff requires not just technical skills but soft skills. We hear lots of talk about the importance of soft skills. They're way more important than I ever imagined. You have to speak in terms of other people's interests. What's in it for them?

Speaking of soft skills, can I specifically recommend Dale Carnegie training? Specifically, the course in human relations. The introductory course. It's a game changer. It literally changed my career and my life. Dale Carnegie training is about much, much more than speaking in public. If you can't get your employer to subsidize it, pay for it yourself. I promise you, if you don't do so today, a year from now you'll wish you had!

Mala: What's the role that documentation plays in your job? I think you mentioned a little bit of it before.

Jimmy: I have a multifaceted answer here as well. First of all, good documentation, whether it's good commenting in code or formal user guidance, is a function of engineering discipline—period.

I've never been a believer in job security in terms of keeping things close to your chest. In fact, that's short-sighted, a function of a scarcity mentality and insecurity—a lack of confidence I want no part of. This attitude is not only self-destructive, it's toxic for teams. I have always been a person who shared their knowledge. Share the love, baby! I expect my teammates to do the same. It's a team culture thing. Oh, that reminds me. If we don't get to this in a subsequent question, I'm going to circle back about playing nice on a team. Anyway, as I said, good documentation is a function of engineering discipline. Again, whether it's code comments, you have an internal repository for your team documentation, your help desk, and end user guidance. For me, time for good documentation needs to be written into the spec for project scoping. It cannot be an afterthought if you want it to be useful, usable, coherent, lucid, etc.

I'm a big fan of sharing my learnings, not just within my team, but also, in terms of external consumption—for example, white papers, presentations, etc. My blog was on the top six or seven percent in Microsoft for many years in a row when I was working as a consultant on MCS, at SQL CAT, and then at MSIT. Even now, years after leaving Microsoft, it remains in the top ten percent, which can't possibly last for much longer because much of the content is becoming less and less relevant pretty darn quickly! But yes, I'm a big fan of documentation. Good, comprehensive, coherent, actionable documentation. And again, it cannot be an afterthought.

Mala: What are your favorite books, blogs, avenues of learning? I know you mentioned certification is a big one. Other than that.

Jimmy: Oh, huge. Not just Microsoft certification. As you know, I'm a Microsoft Certified Master—one of about one hundred in the world for SQL Server. I was already at the top of my game when I was recruited into that training, and MCM training was icing on the cake. *It was amazing.* At that time the training was on-site in Redmond. It was taught by incredible world-class SMEs, the cream-of-the-crop, you're there face-to-face with people like Paul Randal, Kimberly Tripp, Kalen Delaney, Joe Sack, Adam Machanic, Greg Low, a handful of other people. You're face-to-face learning from them in a focused environment. And the peers, the people we're in class with, these are world-class SMEs literally from around the world. We're in this crazy, super-intense, sleep-deprived environment, stuffing our brains from morning to midnight. It was a once-in-a-lifetime experience, and it was awesome.

Also, I've long been a believer in paying for my own certification. Early on in my career, I made a conscious decision. I thought getting an MCSE would enhance my DBA skills. And the extent to which that was true is shocking even to me. It was an amazing enhancement. It paid for it myself. To accelerate my training I invested in a program that met one night each week for nearly a year. It was ten grand—out of pocket. Some of my peers at the company I was then with—you've heard this broken refrain, too, I'm sure— said, "Well, if my company's not going to pay for my training, then I'm not going to." That is *exactly* the wrong attitude. Economist Walter Williams says that there is no better investment you can make than in yourself. Believe me, if you knew about my stock picks, you'd agree! Dr. Williams is right. Investing in myself—in my own education—*is* the best investment I've ever made. Again and again and again, it's paid off. It's been incredible. That's been huge throughout my career.

My favorite books are the *Inside SQL Server* series. I remember early on in my career I got Kalen Delaney's *Inside SQL Server 2000* [Microsoft Press, 2000]. I never slept alone. That book was with me every night—annotated, bookmarked, dog eared, highlighted from cover to cover. I read it multiple times. It made me a much better DBA. I don't read books like I used to anymore. Sadly, the time just isn't there. But I love Paul and Kim's newsletter. Those guys are the best. My most recent nugget is MVP Kendra Little's new podcast. It's awesome. The Dear SQL DBA podcast.[3]

Brent Ozar has so many great offerings. His office hours, his blog, his blitz scripts, etc. There are so many others. I'm sure I've left out dozens of great names.

---

[3]https://sqlworkbooks.com/dear-sql-dba/

Mala: What are your recommended ways of stress management and developing healthy work/life balance?

Jimmy: What a great topic! For a long time, Mala, I was unsuccessful at this. I devoted *everything* to my career. My health suffered for it. As you may know, I have gained and lost eighty-five pounds not once, not twice, but *three different times*. Call me a slow learner. Yo-yoing up-and-down like that is nuts. Today, I make fitness a priority. I have been at or near my goal weight for nearly six years. I'm proactive by scheduling my fitness activities—they're not afterthoughts. Simply, fitness is a lifestyle I've chosen. Whenever I have projects or jobs, etc. I make sure my team is aware of my priorities. I have a partner—my fiancée, Allison—she and I make it a point to work out together five to seven times each week. Our busy schedule is challenging, but it's great together time. There are a few things more satisfying than to have her beside me as we ski, hike, bike, do yoga, do plyo, whatever. Again, if it's not on my schedule, it doesn't happen, so I integrate my sessions into my calendar. Skiing is my fave. I moved to Utah for the mountains.

I do thirty-day challenges with a couple of Microsoft friends, fellow MCM Jens Suessmeyer and architect Steven Wort. We're in our sixth round. Nearly every day I do a six-and-a-half to seven-minute plank. A couple of years ago while planking I would listen to a mix of *Uptown Funk*, wondering whether I'd ever get to the end of the song. It's "only" five-and-a-half minutes, so now it's much too short.

Allison and I not only workout together, we cook together and we eat together. For example, she introduced me to some healthy choices which heretofore I'd ignored. Every morning for breakfast I used to eat #bacon literally every day—like a good SQL DBA. Now I have bacon maybe once a month and that's an accident. My practice formerly was to eat meat at each meal. Today I just don't get around to it like I used to, and beef is a rarity. Spinach is a super food. I put it in or on top of nearly everything. I eat a lot more beans than I used to. Here are some words I live by: You can't outrun your fork. Abs are made in the kitchen

Mala, I need to mention something else. Not too long ago I sat a ten-day silent meditation retreat. It was yet another game-changer for me. I learned so much from it—the power of equanimity. The awareness of constant change. How to detach and observe the world around us and the emotions inside us, and so very, very much more. In fact, the tools I learned in meditation provided quantum leaps to my planking—added a minute overnight—and spinning— instantly adding eleven percent to my PR [personal record]. The things we hear about so many of our challenges being mental—it's so true. As they say, "When you're in your own mind, you're behind enemy lines." Be clear, meditation isn't merely a tool for fitness, the attitude adjustment it provides improves every aspect of my well-being. I can't recommend more strongly that you find a Vipassana retreat at your earliest opportunity—the life you improve may be your own.

What else do I do to relax? Well, there's *Game of Thrones*, of course. Who amongst us don't think to ourselves, "The night is dark and full of terrors," and "What is dead may never die." And I don't even drink, but I howled aloud when Tyrion said, "I drink, and I know things."

And *West World*—the uncanny valley is uncanny indeed. "Have you ever seen anything so full of splendor?" Like I said earlier about AI, we'll make great pets.

Mala: What's your style of interviewing a data professional? What do you look for in somebody you want to hire?

Jimmy: I always start out simple. No matter what a candidate says they can do or have done. I start out with simple things such as select statement. I present a scenario where we build on that select statement to exercise various clauses. Simple select, add a filter or two, add a group by, add a join, include a subquery. I see if there's familiarity with any hints, for better or for worse, and things such as that. Then I talk about schema. I'll probe to confirm that they've represented their skill sets honestly. Integrity and trustworthiness are important to me, as is trainability. Engineering discipline, for me, covers a lot of different things, not just documentation. I get a feel for that. Also, I get a feel for flexibility and friendliness. Not just being a smart person, but the ability to get along. Not just to be a nice person, but how good a fit will they be in the team. I've worked with some very bright people in my career. And someone who's bright, open-minded, and empathic? Bring 'em my way!

I dislike working with 'my way or the highway' kind of people—the kind who insist it has to be done this way or that way—their way. Sometimes there *is* only one way to do it right. Almost always there are more. We have to be flexible. Technical skill is critical, but only one important facet. I certainly want somebody who hasn't misrepresented themselves. Somebody who's able to say "I don't know" without embarrassment. Good people skills in general and someone who's a good fit on the team. That's what I'd say. All right?

Mala: Why do you recommend people be involved with community?

Jimmy: Oh, Mala, you know as well as I do what #SQLFamily is all about! Amongst the great honors of my life—albeit the most melancholy—was my having been invited to speak this year at two eulogies for #SQLFamily.

Tom Roush and Robert Davis were two of my best friends. They both left us this year. When they died, their lovely brides—Cindy and Chrissy, respectively—asked me to speak at their memorials. I flew to Seattle to do so.

I spoke at Tom's along with Mike Walsh and Yanni Robel. Several others spoke testimonials at Robert's memorial—Kendra Little's words were especially moving; Dale Hirt, Michael DeCuir, Dan Brennan spoke of Robert in likewise poignant fashion. It was a real honor—again, the greatest honors of my life.

My involvement with community has been a huge asset to my career. I have started two user groups back in Indianapolis—IWUG and IndyPASS. As part of that work, I became involved directly with Microsoft. A young yet not very knowledgeable DBA can do a lot worse than work side-by-side with well-respected Microsoft personnel! That experience helped me get my foot in many doors. What an honor is it year after year, Mala, when I submit to SQL PASS, and I get selected.

One thing that's important is to be willing to pay for your conferences—including travel. Even as a Microsoft employee, I paid my way to PASS out-of-pocket at least twice—using vacation, paying for the conference pass, airfare, hotel, and food.

I knew my Microsoft manager was not going to send me to SQLBits, so I took a vacation and flew to London. I submitted three sessions hoping one would be selected. I was shocked when all three were selected. I got to fly across the pond fully jet lagged to deliver them. Be careful what you ask for!

Likewise for SQLSaturday. I've spoken at dozens—over fifty—the vast majority for which I've paid out-of-pocket.

In Dan Millman's book, *Way of the Peaceful Warrior* [HJ Kramer, 2006], Socrates says, "There is no higher calling than service." And it *is* immensely gratifying to give back to the community. Mala, I was so proud of you when you received the PASSion Award from the Professional Association for SQL Server at the Summit!

I've spoken on four continents including the largest conferences on three continents, including eight different times at the PASS International Summit. I'm going on six years now in a row, including four years with multiple sessions.

It's all a function of getting my toes in the door in these user groups over a decade ago.

Mala: That's fantastic.

Jimmy: It is. It's pretty incredible. It's not that I was especially skilled, but it's my passion and my enthusiasm and a little bit of technical skill to have combined to provide these opportunities. I remain amazed what that a small town boy such as myself with no prior computing experience has been able to do! From "aspiring geek" to enterprise architect at Microsoft, ExactTarget, Fusion-io, SanDisk, and Western Digital. So, bottom line: the SQL Community is incredible. Take advantage of it!

Mala: Describe an interesting story you'd like to share. It does not have to be work related.

Jimmy: Let me give you three quick ones. You've probably heard of Richard Simmons. He's the guy who is responsible for selling the world's largest selling exercise video of all time. Have you heard of *Sweatin' to the Oldies?*

Mala: No.

Jimmy: Okay, well, Richard did a series of exercise videos. I think four, or five, or six of them. I was one of the stars in his very first original *Sweatin' to the Oldies.*

This was way back in the early to mid-eighties, at a time when Jane Fonda's "Burn, baby, burn!" was popular. He outsold Jane Fonda by a large margin. Anyway, this was before my geekly career. I was living in LA. I worked out at Richard's Beverly Hills studio. And I was the token straight male in his first video. Seriously!

Here's another anecdote. While out there on the Left Coast, I was engaged to a lady named Evelyn Guerrero. She was the female lead in several of Cheech and Chong's movies. She was a Playboy model. That didn't work out. She ended up marrying Pat Morita, the guy in *The Karate Kid* movies.

Lastly, I had a job as an administrative assistant for the CEO of what is now Samsonite. Amongst my enviable duties was driving his Porsche 928 and Jag—and coolest of all was picking up a Ferrari from the airport. This was way before TSA and high security. I picked up his shiny new Mondial Cabriolet, put the top down, and drove it straight off the tarmac—first onto Century Boulevard then the 405—where on the way home to Beverly Hills, I hit well over 120 miles per hour. Not too long afterward, I totaled the Porsche—I did a 540-degree spin on Westwood Boulevard right behind UCLA. It was fun while it lasted...

Key Takeaways

- Specialization—not generalization—is a key to success. In a former life, a mentor admonished me to be the best guy in the room at something. Be that guy (or gal).

- Don't be afraid to lose your job. Have the skills and the confidence to be willing to walk away if things don't work out. It *always* leads to something bigger and better!

- Be proactive yet be patient. Change can take a year or more. It is not realistic to talk about vast storage and other infrastructure improvements in the context of overnight changes.

- Master soft skills, including public speaking. Form collaborations. Learn how to get to yes. Instead of looking at something as impossible, reframe your point-of-view and ask, "How can I make it so?" Master fearlessness. Remember, you are a ghost driving a meat-coated skeleton made from stardust riding a rock flying through space. Fear nothing.

- Prioritize your health—physical and mental. Consider sitting at a Vipassana meditation retreat.

- There is no better investment than in yourself. Again and again and again, it's paid off. Consider paying for your own training and conferences. And don't look back!

Drew Furgiuele

Senior Database Administrator, IGS

Drew Furgiuele *is a senior DBA that lives in Dublin, Ohio, who is passionate about SQL Server and PowerShell. When he's not accidentally dropping tables in production, he likes writing automation scripts, blogging about SQL Server replication, wiring electronics, playing board games, and spending time with his dog. He's also not embarrassed by his Spotify playlists. He blogs on his personal website (http://www.port1433.com) and spends most of his daily life tweeting about dogs.*

Drew speaks locally in the greater Columbus, Ohio, technical community, as well as nationally and internationally. He's a contributor to the great open source PowerShell module dbatools.io, which provide (as of the time of this writing) more than 300 SQL Server administration, best practice, and migration commands.

Drew loves crazy ideas, such as the High-Altitude SQL Server Project (HASSP), in which he's attempting to put a running SQL Server instance into space. He can be reached at twitter as @pittfurg.

Mala Mahadevan: Describe your journey into the data profession.

Drew Furgiuele: For the longest time, I thought I wanted to be a developer.

When I was still in high school, I took some advanced placement [AP] courses that were provided by the University of Pittsburgh and were centered around coding. We did projects in languages like FORTRAN, Pascal, and C++. I wasn't great at it, but I wasn't bad at it either and I accrued a fair bit of credit hours in the years before college. When the time came to apply, I only really wanted to go one place. Pitt. And I wanted to major in computer science.

My first semester went well. Aside from the required courses and electives I took, I also enrolled in several of the beginning classes for the computer science degree program. Things started out okay. I was a straight "B" student in pretty much every course and a lot of the things we worked on were things I saw in the high school AP classes. Once the second semester began though, it became clear very quickly that I was probably not cut out to be a developer.

I remember the exact moment it happened. We had a class lab assignment to design a clock interface. The program, which was to be written in Visual Basic, had to read the system time and display an analog clock face on the screen, and the hour, minute, and second hands needed to be animated. If you think of an analog clock face, you already have a strong idea of what a functional analog clock face looks like. there's probably a circle with three lines of differing lengths that start in the middle of the circle and extend towards to the edge, and there are also probably some sort of marks or numbers around the perimeter that, when the different lines or hands of the clock point to it, tell you the time. If you've ever worn a watch, you know what this looks like.

Programming something like this, though, was so far outside of anything I've ever done before for a couple reasons. First, this was a graphical program. Every exercise, lab, and final I had ever taken up to this point was in theory or "practical" code samples. Not only was this a graphical program, but to render this clock, you needed to know math. Higher-level math, like trigonometry, to help determine what point on the circle a line should point to. And that's to do it once. How about animating it and making it update properly with the passage of time?

I was scared. I have never been good at math. I was already sweating the fact that I would need several semesters of calculus as required courses in this degree program. But I always assumed I'd deal with that when I'd get to it. Here I was in the second semester of a 100-level computer science course, and already my lack of knowledge, and honestly, desire to learn advanced mathematics was jeopardizing the very thing I always wanted to do.

I did finish the exercise. My clock face was more oval than a circle. The second hand of the clock worked great for the first fifteen seconds, and once the hand passed ninety degrees things got weird. The program also had a tendency to crash if I ran it on Sundays. I still don't know how I accomplished that one. Needless to say, I didn't score well on this exercise. Worst of all, it left me really defeated.

Shortly thereafter, I decided that maybe I wasn't cut out to be a developer.

If that seems like a harsh decision based on just one exercise I couldn't do, it's hard to argue with you. Truthfully, I was already sort of feeling like being a developer wasn't something I would genuinely enjoy doing for the rest of my life. College just helped reinforce that. I still loved computers, and coding was fun if it was on my terms. I was struggling to figure out what sort of degree I wanted, but fortunately, I wasn't alone.

A friend of mine down the hall in my dormitory was in the same boat, and he told me about a different degree program offered by the university. Information science. It still had some programming requirements and courses to pass, but it also had things like "Human-Computer Interaction" and "Information Storage and Retrieval" courses. It even had some web-development specific courses. I changed degrees and started down that path.

Four years later, after graduating from the University of Pittsburgh's School of Information Science, I started my first full-time job as a help desk technician at a pharmacy company in Columbus, Ohio. My boss at the time realized I had a degree and wanted to give me a chance to use it, so aside from lugging printers up flights of stairs, wiring up Unisys and IBM terminals, and resetting passwords I was also working off-hours developing internal applications.

It didn't take long for me to start developing applications that needed persistent storage, so that's when I got my hands on my first production relational database. Microsoft SQL Server 2000. The company had a huge investment in IBM DB2 and some Oracle database servers, but I couldn't get access to those to stand up new databases. I was relegated to this "other" database platform that didn't have any official support from corporate IT, but it turned out I didn't need it. At least not at first.

I fully embraced my role as the accidental DBA. I made all kinds of mistakes and learned several lessons about database administration the hard way. As I continued to develop applications, my focus shifted from actually creating new applications to supporting the data they were collecting. I had to learn how to write complex queries, how to do performance tuning, and even how to cleanse data. Most of all though, I got to spend time with my end users, learning about how they consumed the data being collected in the databases I was now maintaining. That helped me learn how to create effective data structures to best store and eventually retrieve data.

I did this for many years at my last company, but then I was offered a position where I could manage a team of analysts who would do what I was doing, and I could focus on bigger picture projects.

Management. At first, I thought it was exactly where I wanted to be. I was still pretty young, at a very large company, and I was doing well. This was a great opportunity to grow and the sky seemed like the limit. But then a funny thing happened. I really, really missed working directly with databases. I realized I wasn't happy, and I decided I was time to get back into being a specialized worker bee than a leader. That kind of job didn't exist where I was, so I had to go out looking. It took a few months, but when the right opportunity came along, I jumped and haven't ever really looked back.

Now, I'm several years into my role as a senior DBA, and I know for certain it was the right choice to make. I'm working at a fantastic company that helps me just as much as I help it, and I work with some of the smartest people around that are always challenging me to be better and to learn new skills.

Mala: Describe a few things you wish you knew when you started your career that you know now and would recommend newcomers to this line of work know.

Drew: My first rule of advice, for anything really, is to act like you've been there before. In this line of work, people will expect you to be able to say things like, "Our data is safe." Not only do you need to have the skills to back that kind of statement up, but you also need to exude enough confidence to assuage the concerns of your business partners and leaders.

If you want people to respect your opinion, prove it through your actions and passion. Don't preach to people about what they should do. Instead, try to coach and be a cheerleader for what you think it right.

Mala: What is a typical day in your life as a professional?

Drew: In my current role, my "day job" is that of a production SQL Server DBA. That entails all sorts of things. I should make sure crucial database maintenance tasks—like successful backups and database integrity checks—are completed and audited. I also must monitor our systems for any signs of performance issues using a variety of third-party monitoring tools and a few homegrown solutions as well. I also work really closely with our development teams in both troubleshooting underperforming database queries and also lending a hand and/or an opinion on upcoming projects or refactoring.

Mala: What are some industry trends in this line of work that you are watching out for?

Drew: While I think the death of the DBA role is pretty exaggerated, I do think our jobs are changing and we're being asked to help create solutions, especially around automation, for deploying database code changes in an increasingly agile environment. That also means learning more about topics like continuous delivery and integration. Learning how our roles will be impacted by the continued adoption of containers is certainly something more people are trying to understand, as well as how we can move our data around to new places, like data lakes and multi-model databases.

For me, this all ties back to that my biggest challenge is getting and staying up to speed on cloud-based technologies, and I don't mean just traditional relational databases. Understanding how applications and data live and survive—and even die—away from you, and how the resources that drive those things are provisioned and decommissioned, and ultimately, what they cost are things that seems to always be in flux.

Mala: You are a PowerShell guru. Describe some of your experiences with learning PowerShell, helping the community with free scripts, and so on.

Drew: PowerShell was something I had to learn out of necessity. When I started in my current job, I came in day one and saw all the automated tasks— system configuration to software deployments and almost everything in between, and I was completely floored. Never once in my interview was that even a topic that came up. I still don't know if it was just a function of DBA's traditionally not working with PowerShell, or it just wasn't something I was expected to use much of. However, I couldn't help but feel very intimidated in this new role where PowerShell had such a large company footprint. Maybe I wasn't totally expected to know it, but all our database code deployments were being handled with PowerShell, so I had some incentive to get up to speed.

I had some fantastic peer resources, and I could look at scripts and functions that they had developed as a good place to pick things up. I still didn't make a good connection about how it could help me as a DBA though. There were so many good cmdlets and examples for system administration-based tasks like dealing with the Windows file system or interacting with other software technologies like Active Directory, but SQL Server didn't quite have the tight integration that I was expecting.

That's why you see so many different examples of how to do different SQL Server related tasks with PowerShell, even today. Thanks to a very supportive group of people at Microsoft that work on the "Tools Team" we have much better support for accomplishing automation with SQL Server and PowerShell than we ever have had. I love coming up with automation solutions for daily DBA-type tasks, but the great thing about PowerShell is that it can be used to expand your role to other tasks.

Mala: Describe the HASSP project. A lot of people don't know about it. I certainly didn't until I caught it on Twitter.

Drew: Sometimes I get bored, and when I get bored my mind really starts to wander to some strange places. One day at work I was reading about hobbyists launching high altitude balloons [HABs] and doing things like putting cameras on them and recording the flights. I watched a couple videos and my interest was really piqued. After doing a little more research, I discovered that doing a HAB project isn't overly complex, or even very expensive. Anyone— from seventh-grade science classes, scout troops, and even IT professionals all over the world—was launching them. As I was reading about them, I had a crazy idea. What if we attached a database server to a balloon and let it fly? Would that even be possible? If it was, what sort of technical challenges would we need to solve? What we would store in a database at 100,000 feet up, and how would we even collect data the data that we wanted to store?

Right then and there, I knew that I wanted to try. I stood up at my desk and announced loudly to whoever could hear me—which turned out to be almost everyone, I didn't realize I was shouting at the time—that I was looking for volunteers to join a project team that would "put a database in space." A lot of people laughed it off.

But not everyone. I had a few coworkers say they wanted to be a part of it, so I formed a "project team" that set aside time at lunch and after work to help bring this project to light. I acted as a kind of "project manager" for the team and helped assign different tasks to different people. For example, we had people who worked on figuring out the best cameras to use, we had people who helped find and test different system boards that might be able to run SQL Server and still fit in a small form factor. A couple team members also wrote software that was responsible for collecting and writing data to SQL Server. We had people who solely focused on coming up with a container to house all our sensors, batteries, and server. I still helped with some of the technical stuff. To track the balloon in flight, I had to use a technology that required an amateur radio license, which I didn't have, so I had to study, apply, and pass an exam. I also was responsible for coming up with the sensor wiring and testing that would ultimately be the collection point of all the data. We had components that could measure environmental and atmospheric data, like air temperature and pressure, and we also had an accelerometer inside to see what kind of forces would be interacting with it in-flight. Finally, I also put a GPS receiver inside the package as well, to give us better position and altitude readings.

It took us about four months to come up with a working prototype, and then we began testing. We ran the server fully enclosed in different scenarios. We tested our components using dry ice to simulate very cold temperatures, and we even tied the entire payload to a drone and flew it around to see what would happen. Everything was looking good, and we scheduled our first launch after finding and paying for a balloon that could hold enough helium to give us the lift we would need to, you know, fly.

We did all we thought we could and set out for our first flight. There's a video of the attempt online, but the short version is we failed. The technical reason we failed was "failure to properly account for wind speed and accurate measurement of required vs. actual balloon lift" or if that's too much, you could go with "we hit some power lines."

The team and I were crushed, but we weren't about to give up. We sat down and had a frank discussion about what we did well, and what we could have done better. It became apparent quickly that we were so focused on building a server that we forget that we were building a server designed to fly at 100,000 feet in the air and we didn't do enough testing or research into what it actually takes to fly a balloon to that altitude. We all agreed that our main point of failure was rushing to launch in sub-optimal conditions and our inability to accurately measure the lifting force we were generating with the amount of helium we were using.

After talking it over with the team, we came up with a few simple tweaks. First, we decided to work indoors for almost the entire pre-flight operation. By doing this, we would eliminate the external factors and stress in trying to get everything ready to launch outside and trying to keep a balloon with a seven-foot diameter from flying away before you want it to. The next thing we determined was that we were really bad at measuring the amount of lift our balloon is generating, so we secured a fish scale and sort of used it in reverse. Instead of something pulling down to measure how much it weighs, we turned it upside down and had it show how much force our balloon was pulling with.

Those two tweaks are what it took to get us the final pieces we needed to attempt our second launch. And yes, we did manage to fly our payload up to almost 100,000 feet and we successfully recovered it when it came back down. We were also able to track it for the entire flight, and we even had our cameras record the flight as well… except our SQL Server shut down about thirty-five minutes into the flight. So that means we need to try again, and the team and I are fully committed to seeing it through.

While the project was fun, it had some other "unstated goals." The first is that there was a larger lesson that showed a lot of corollaries in our day jobs as IT professionals. Things like missed requirements, or focusing on the wrong parts of a project, to insufficient testing. It's easy to keep falling into those same types of traps or patterns repeatedly, and I feel as if coming up with ways to both mitigate those risks and having a process in place to deal with them if/when they happen is key. It was also a great opportunity to interact and share our journey with the SQL Server community, both by keeping a running diary of the project on my blog and social media and by setting up live streams for people to watch the launch and recovery.

Finally, and maybe most importantly, it was a great way to spend some time with my coworkers and friends. This project helped bring us, with all our different domains of knowledge, together with a common goal that we all got to contribute to. I'm fortunate to work at a place that not only fosters that kind of relationship among peers but encourages people like me to find ways to continue to foster it. You might work at a place that does too, but you won't know till you try to do something like this.

Mala: What is your experience with agile methodologies with regards to database administration?

Drew: That's one area I had to get up to speed in my current role very quickly. I had come from a place that had very glacial software releases, with usually one large release every other month or so and smaller, hot-fix type releases in between. Some projects and teams, like mine, had a little more liberty in how we handled deployments so we moved somewhat faster.

Once I started my current job though, that all changed. Suddenly, I was part of an agile team, even when I was on the "infrastructure" side of the IT department. I had never had to deal with things like sprints and iterations, CI and QA, and even the concept of formal, scripted releases. The thing that tied all this together for me, though, was automation especially with using PowerShell to deploy database changes and updates automatically. So instead of having to manually deploy these changes, my focus instead has shifted to knowing what's going to be deployed and helping development teams overcome problems they might have with them. And finally, I think when you talk about being agile, there's always a continuous improvement aspect to how you practice that methodology and strive to make it better and faster and less of a burden for everyone involved.

Mala: Describe your experience with cloud adoption.

Drew: Honestly, I haven't had to work too much with the cloud, but even as I answer this question this is changing. In my current role, I've got a couple non-production database instances running in Azure, as we explore the platform and what it gives us and what it doesn't. We're also working with cloud providers like Microsoft and Amazon to explore leveraging them for something like disaster recovery. The technology is fascinating, but I'm not so much concerned with how they work, and instead more focused on how I interact with them. Again, being such a fan of PowerShell is sort of crucial here because you can interact and automate a lot of cloud-based deployments to minimize the amount of time you spend provisioning and decommissioning resources and instead let you focus on bigger-picture items like architecture.

Mala: What are some of your favorite tools and techniques?

Drew: My absolute favorite tool to have is a repository of tools. I use things like Google Drive and Microsoft OneDrive to store collections of handy scripts I've collected or written over the years. There are so many times I need to re-run something or recall something I once did that just having my own personal repository keeps me sane. I don't make my collection publicly available—like, say, on GitHub or any other online code or file sharing service, but I have shared things with people from time to time.

I also haven't had access to enterprise-grade database monitoring tools for most of my career, but now I do and I can say that they really show their value quickly in helping a DBA quickly identify areas of concern. Some people may only look at the price tag of a tool, but if you want to balance what a piece of software like that costs versus the amount of time it might take a DBA like to me aggregate and digest all the information these tools so easily roll-up for analysis, I think you might find it pays for itself much faster than you realize.

There's also a lot of great tools out there too that are free. I love tools like SentryOne's Plan Explorer for helping tune queries, and I also like Brent Ozar's free tools to diagnosing server slowdowns and configuration issues. Finally, I would be totally remiss to not recommend the excellent dba tools open source PowerShell project to help you automate and work with SQL Server instances with PowerShell. You can't find a harder working and innovative team helping to make PowerShell support for SQL Server what it always should have been.

Mala: What sort of struggles have you faced with managers/management or business side of things, and what is your recommended approach to handling that?

Drew: I've been fortunate that I've almost always had great, supportive managers that I've resorted to. Some managers have been familiar with the roles and responsibilities a DBA has, and some didn't know much at all. Neither of those scenarios is good, or even bad. It just shaped how I had to interact with them to express what I was working on, or why I wanted to approach a problem or a project in a particular way. The respect I have for my managers and the respect that they show me has always been mutually earned.

The bigger conflicts I've had in my career are with other teams, specifically any team that manages hardware or budgets. It can be hard—for me, at least—to properly justify why I need something and/or why the thing I need costs as much as it does. For instance, in my first job I just could not get approval to buy an enterprise monitoring tool. I certainly knew the value of having one and what the benefits would be, but I did a bad job of explaining the need to my boss and the finance and purchasing people. It wasn't that I couldn't always communicate that I needed to spend money on something. If a project I was working on needed something like a new server or a new license for SQL Server, those are a little easier to justify because without them, there simply would have been no project. Quality of life and peace of mind purchases were always—and still are, sometimes—tougher to quantify and eventually justify.

Mala: What sorts of issues are caused when you interact with other technologists? Which ones—developers, SAN admins, network folks—are you most likely to clash with and why?

Drew: Along with what I said to the previous question, the same thing goes for configuration changes to server hardware, storage changes, or even network configuration. I know I want to live in a world where I never have an underpowered SQL instance in production, but it happens. Some things just need to be scaled with hardware over time. It's one thing to say I need to spend money on hardware or upping the amount of money I spend cloud infrastructure, but in today's world of virtualized servers, even requesting a modest boost to resources can become a protracted argument over what's really needed.

At first, I would take a pretty firm stance on requests, but I've found that partnering with the folks that are responsible for these resources instead of demanding things from them seems to go a lot farther. And, it's hypocritical of me to get mad when someone doesn't give me everything how I want it, and when I want it when I can't say I bow to each user's requests every time. It's a two-way street to effectively and respectfully work with those teams.

Of course, if all else fails, data helps. Be prepared to make your case and argue for it with data that backs it up. Decide on some metrics to collect, measure them over time, and use them to support your argument. And then, once you do, if that helps enact your desired changes, don't stop there. Collect data after the fact to show the new benefits your changes made. Not only does this help drive home your requests' purposes, but it also helps these other teams see the value they add to your world.

Mala: What is the role of documentation in your job?

Drew: I don't think anyone is going to say documentation isn't an important part of anyone's IT career, and I'm no different. It's important to properly document your processes. This isn't—or shouldn't be—news to anyone. The bigger challenge is attempting to standardize and aggregate all the different pieces of it. Whether it's something like database entity-relationship diagrams, best practices, or even who to contact if something breaks, having all that stuff is written down is good, but having it centralized and accessible is even better. That's why I think it's important to not just have good documentation but organizing it in an internal wiki or repository of some kind is just as important.

Mala: What are your favorite books, blogs, and other means of learning?

Drew: I learn best from examples. It can be blogs, webcasts, videos, even books. I just want the code to play with. I really enjoy very deep technical dives into subjects that go beyond general overviews and instead focus on practical examples. I also feel like presenters and authors that focus more on demos tend to keep my attention and help me learn, and it's something I emphasize in my own presentations as well.

I also crave examples of efficiency. Once I learn how to accomplish something with technology, I want to know if there's a faster, more efficient way to do the same thing. This was a habit I picked up once I really started working automation like PowerShell, because once something is automated, I'm not satisfied that it's just automated. I want my automation to be fast and simple. I like to find examples of ways to shed complexity too.

Mala: What are your recommended ways of stress management and developing healthy work/life balance? How do you handle long grinds on the job?

Drew: Hobbies are what help keep me sane. I'm a big board game fanatic, and I've taken it upon myself to set up quick gaming sessions during lunch breaks at work. We play a quick game—lasting between fifteen to forty-five minutes on average—to help provide a nice break for myself and coworkers. Most games only allow four or five players at a time, so I also try to find games that support larger player counts if more people want to join. These can turn into great social engagements I feel like not only does it provide great stress relief and some brevity, but it also helps reinforce some teambuilding as well.

Outside of that, I think keeping fit is a big personal priority for me. About five years ago, I discovered CrossFit and I have never been in better shape in my life. When I say shape, I don't mean just strong. I've been able to lose weight and keep it off, increase my mobility and flexibility, and stamina. It also helps that I joined a very supportive gym community that has a lot of vested interest in helping their members be the best they can be. I've never done well joining a large gym or even a local community gym. Just getting to the gym after a long day can immediately make me feel better because I'm in such a positive space with amazing people who want to help push me. I can do things now that I never thought I could, like free-standing handstands.

I'd be lying if there weren't days I basically have to drag myself there, but I'm never sorry I do. I know that maybe my experience is somewhat anecdotal there, but I encourage everyone to try to find some way to fit a workout into their weekly routines, even something small. Getting started is often the biggest hurdle.

Mala: How do you work long hours that are boring and not challenging whiles still keeping motivated?

Drew: "The grind" can be real at times. There have been plenty of occasions that I've had to work extra hours and weekends that make me question why I do the job that I do, but I think some of that just sort of comes with the territory. I don't have to do it super often anymore, but there was a time at my current job where we were on a project that required a lot of dedicated weekends to convert a bunch of applications and data to new systems and that was just the only time we could carve out to do it. It helped that I wasn't alone in the effort it required, but I was really putting in a lot of hours over weekends, especially overnight. So not only was I gone for most of a weekend, I would start the next week exhausted.

What kept me going though was knowing that the work I was doing was important to the company. Even though it was mostly babysitting some processes punctuated with configuration changes here and there, what was tedious and mostly automated work to me was of vital importance to the overall goals of the company and the satisfaction of our customers. Just because I found it boring didn't mean it wasn't critical. When I viewed my work through that lens, the fog of boredom was lifted and I became a lot more focused and engaged.

Still, there may be times where I question why I'm doing something the way I'm doing it. To conquer that kind of slumps, I must switch gears and instead of just focusing on completing a task or project, finding a better way to complete the task or project. If I have the leeway to do something that—and even if I don't at times, I'll take the opportunity to do it because chances are I'm going to learn something new and exciting.

Mala: Describe your style of interviewing a data professional. What do you look for, and what are some examples of questions you ask?

Drew: For me, it isn't so much a question of, "Can you do the job I'm interviewing you for?" It's more, "Can you be a person that's willing to learn and not be toxic?" Fitting into an existing team means that anyone coming into a new role doesn't have to know everything "by the book" but instead be willing to contribute to the team. That means finding "best fit" individuals who I feel are going to get along well with whoever they are going to be working with and the customers they'll be serving.

Skills like effective communication and individual passion for the job leave a bigger impression on me than on "I can tell you every isolation level and what they do," or "I know how to write a complex windowing function from memory." I can always go online to find an answer the latter, and I'd rather that the former come through in an interview. To that end, I'll spend less time talking about roles and responsibilities and instead try to just talk about things a candidate is passionate about. I want to get a feel for their motivation and demeanor, and just having a fifteen or twenty-minute conversation about their interests and hobbies helps that come through.

Mala: What are your contributions to the community, and why do you recommend people be involved with the community?

Drew: I first started contributing to the SQL Server Community almost four years ago when I started speaking at local user groups and my first-ever SQLSaturday. Since then, I try to commit to at least a half-dozen speaking events a year in and around the Microsoft Data Platform. In addition, I try to blog twice or more a month, and take part in online webinars and doing training events remotely. It's been a very rewarding experience to be so involved, and I've been able to network and pick the brains of some of the smartest people working in my field.

I think everyone should be involved in sharing knowledge, regardless of how it's done. You could start speaking, you could blog, or you could record yourself doing something funny or irreverent. Just share! You don't have to have a new topic, or something someone else hasn't covered before, just your perspective. Not only will you more than likely help someone who might have a problem you know how to solve, sharing your skills and knowledge is a great way to practice and keep your own skills sharp.

Mala: If you had one superpower, what would it be and why? It does not have to be work related.

Drew: I've never felt I've been very creative. I am not good at expressing myself artistically and I have tried many times in my life to learn an instrument. There are probably a lot of reasons I haven't, but it's something I have regretting not putting more effort into before. So, if I could have one superpower I would want the ability to play any instrument. Not necessarily play it flawlessly, just being able to pick up, say, a guitar, and know how to play chords or a few bars from a popular song, or to sit down at a keyboard and tap out a melody. My brain has never been able to comprehend how people can do that, and it's something that I'm very jealous of people who are able to do that. I wish my brain had that kind of creativity wired into it.

Key Takeaways

- My first rule of advice is to act like you've been there before.

- Don't preach to people about what they should do. Instead, try to coach and be a cheerleader for what you think it right.

- Some people may only look at the price tag of a tool, but if you want to balance what a piece of software like that costs versus the amount of time it might take a DBA like to me aggregate and digest all the information these tools so easily roll up for analysis, I think you might find it pays for itself much faster than you realize.

Favorite tools: SentryOne Plan Explorer, Brent Ozar's free tools

Marlon Ribunal

SQL System Support Engineer, JustEnough Software

Marlon Ribunal *is a data professional primarily focused on the Microsoft stack. His work experience includes database administration, SQL development, query and performance tuning, ETL (extract, transform and load), and business intelligence. He is the primary author of SQL Server 2012 Reporting Services Blueprints (Packt Publishing, 2013). He is the technical reviewer of SQL Server 2017 Machine Learning Services with R (Packt Publishing, 2018).*

Marlon's love of continuous learning is leading him toward big data and data science, and he is gearing up for in this next adventure. After ten years in the aerospace industry, Marlon recently joined a software company that caters to the retail industry. He is learning a lot about the complex world of supply chain management, and he is enjoying every minute of it.

Marlon is passionate about SQL Server. He loves learning about the technology. Mostly, he is a learner, but he loves sharing the knowledge he has accumulated. He is an introvert by nature but loves connecting with people through technical community events such as SQLSaturday, social networks, on his blog, and other forums.

© Malathi Mahadevan 2018
M. Mahadevan, *Data Professionals at Work*,
https://doi.org/10.1007/978-1-4842-3967-4_26

Outside of the technology sphere, he is active in Freemasonry in California. He is the current Senior Steward. He is also a member of the Shriner International with the Al Malaikah Shrines in Los Angeles.

You may catch Marlon on Twitter @MarlonRibunal. Learn along with him on his technical blog—SQL, Code, Coffee, etc.—at http://marlonribunal.com.

Mala Mahadevan: Describe your journey into the data profession.

Marlon Ribunal: My journey into the data profession started in a small company in Irvine, California, back in 2002. The primary business of the company was digitizing physical documents for archiving. I was not hired as an IT staff at the beginning but as a production associate whose job included feeding boxes of papers to big scanners, indexing output data in the GUI, and data quality assurance. At some point, I became a production lead—a position that required me to submit production reports to my managers. I taught myself Microsoft Excel and Access to automate my reporting. Microsoft Excel and Access introduced me to the world of data.

After a couple of years on the production floor, I moved to the IT department as an IT support. The backend database of the software suite that we were using to digitize documents at that time was SQL Server. So, shortly after the move, I got involved with the general database support and maintenance. My daily job included, but not limited to, responding to user issues, trouble shooting database connectivity issues, data analysis, reporting and such.

I left that job in 2007 equipped with more knowledge about SQL Server to join a giant aerospace company. My primary job focused on engineering data analysis along with SQL/ETL development, process analysis, and data quality assurance.

After 10 years in that job, I felt I needed a new challenge and, so, I left my job in the aerospace company and joined a software company that offers demand-driven solutions for big retail brands. My day-to-day job now involves making sure that customer applications servers are up and running, responding to production issues as they arise, SQL development, production database administration, and monitoring the health of our customer SQL servers.

Mala: Describe a few things you wish you knew when you started your career that you know now and would recommend newcomers to this line of work know.

Marlon: There is one thing that I wish I knew – that I needed a career plan to get ahead of my career. I may have probably fared better had I planned well for my career. In all the years that I was working on data, I would divert my attention to lots of other things that are not necessarily consistent with my career objectives at that time.

The most important thing in career planning is knowing exactly what you want to achieve. So, the adage "failing to plan is planning to fail" [often attributed to Benjamin Franklin] is not only true to a sense but also true when it comes to one's career. My advice for the newcomers: put together a career plan that you can use to guide your job choices.

If I can do a do-over with my career, I would probably start with a career plan that is designed toward data engineering. I would then list down all the roles of a data engineer and the job description associated with those roles. The next step then would be to look for resources of where I can acquire and how I can learn the necessary skills to meet the requirements to fulfill those jobs.

Another thing that I wish I focused on more earlier in my career is the concept of principles vs technology. It took me a while to decide whether I wanted to focus on Oracle versus SQL Server. I took Oracle fundamental classes while I was working on SQL Server. This went on for years. I did not realize that the technology is less important than principles—that knowing the principles of, for example, database administration is more important than knowing how to manage the database in either Oracle or SQL Server. Equip yourself with principles and techniques instead of the nuances specific to technologies.

Mala: What is a typical day in your life as a professional?

Marlon: As part of our company's global delivery group, a big part of my job is to help ensure that our customers are satisfied with the use of our software products by managing their technical issues and delivering solutions as they tackle their complex demand and supply management challenges with the use of our software.

A typical day in my professional life would include database management routine tasks, such as clearing alerts and notifications in our SQL Server monitoring software, reviewing daily jobs and their performance, and checking that backups are taken at regularly.

I also do troubleshooting of production application issues. In a typical day, issues like slow performance, deadlocks, application timeouts, and various data issues are common. The most critical part in delivering resolutions or solutions is ensuring that we not only deliver them with accuracy but also deliver them within the parameters of the service-level agreement [SLA]. I am also involved in SQL development. From time to time, we receive customization requests or new requirements that are not covered by the current specification of our software implementation. So, part of my job on any given day is writing T-SQL codes or building ETL packages.

Mala: What are some industry trends in this line of work that we need to watch for?

Marlon: My current line of work is aligned with the retail industry. Traditionally, supply and demand are measured by sales figures and historic performance of sales. But with the advent of social media and other technological advances, consumer behavior has somewhat evolved. The process of predicting what products the customers want, when to buy, and where to make that purchase has, therefore, become more complicated as well. Slicing and dicing consumer figures in data cubes are no longer sufficient to come up with an accurate prediction.

Software companies in the retail industry are rapidly embracing machine learning and artificial intelligence techniques as part of their solution offerings to retail brands and harnessing the power of big data in delivering practical and accurate supply and demand prediction capabilities. The need for machine learning and artificial intelligence to power up business intelligence in the supply chain management arena has never been this big.

Mala: Describe a few things that any data professional should know as best practices.

Marlon: If you work with data, you should know both the administration and development aspects of it. Some days you are called to troubleshoot server issues and other days to solve development challenges. As a best practice, a data professional should have both administration and developer skills in their tool belt.

Application developers are well versed in change management as they need to control versions of their software implementation. If you work with SQL development, it is also a best practice to know how to manage SQL codes in the context of change management. Stored procedures, views, tables and other database objects should be managed in a version control system.

And just because something is an established best practice doesn't mean you have to deploy it to your production server without testing. Change in configuration settings must undergo testing processes before they can be deployed to production.

Lastly, all changes must be documented. I don't care if that's a formal full documentation or a simple inline comment in a code. If formal documentation is not possible, leave an intuitive comment that describes the change. It is not only crucial to document the changes but also critical to communicate the same to the people involved in the project. One cannot just over-communicate. Make everyone aware of the changes in email, meetings, online collaboration, and in your ticket system. Again, you cannot just over-communicate when it comes to making changes either in the code or server.

Mala: Describe a few things which any data professional should avoid as worst practices.

Marlon: If you work in any data profession, you are not an island as no man is. You cannot operate as a lone professional in your own silo. You are part of a team even when you are the lone DBA or SQL Developer of the company.

Another worst practice is not committing oneself to continuous learning. Technology is advancing in a rapid phase. You cannot afford to not learn. How far you advance in your career greatly depends on how much effort you are willing to put in learning.

Lastly, be a part of the tech community. Learn and share what you learn in the community. Another worst practice is not sharing what you learn with your team or the tech community. This is critical in your professional development.

Mala: What is your experience with agile methodologies with regards to database administration?

Marlon: I have joined my current employer that is basically an agile software company after ten years working for a company that depends wholly to waterfall philosophy. This was not an easy transition for me. Agile with its iterative approach and the Waterfall with its linear approach are two opposite methodologies.

In an agile environment, the turnaround in every aspect of database administration or development—creating objects, delivering solutions, and such—happens quickly and more frequently. In the perspective of administration that requires keen attention to details, it is not uncommon for situations to go out of hand sometimes. This is why it is very important that you have all your processes aligned with the agile methodologies.

Coming from the standpoint of someone who spent the last ten years of his career in a waterfall shop, the concept of an agile database is quite new to me. I have been learning the critical role of database version control in an agile environment. Every change in objects must be committed first in a branch and only after rigorous testing that the same must be merged in the trunk. With the waterfall, I probably won't have to worry about the branches and trunks until the deployment phase.

The frequent, rapid delivery and the shorter cycle, called a *sprint*, of software delivery in an agile environment demands that the database administration embrace automation. The general administration of the database server is in itself cumbersome. This can easily be made more complicated if you are working with various release versions of the software. Hence, automation is important.

Also, since changes are usually targeted to a specific function or feature of the software at any given time, your approach to database administration must also adapt to this requirement. For example, if there is a performance issue, your mindset should not focus on the overall performance of the database but only to the specific objects affected by the specific module or function having that performance issue.

One of the good things about agile in the aspects of change management is that the process of making changes and delivering the same to the client is quicker and, thus, feedback is quicker to come by. If there are further issues with the deliverable, you can quickly put it back to a sprint and can possibly deliver it in the next release schedule which is usually in a matter of weeks. This is impossible with a waterfall approach in which a simple change request could take months to deliver.

Mala: Describe your experience with cloud adoption.

The biggest advantage of the cloud is the reduction of operating cost and ease of scalability. Provisioning servers from the ground up are costly. The cost of the hardware and the resources to maintain those servers can easily shoot through the roof. Well, you are still spending money if you are in the cloud but you are essentially eliminating the need for additional IT staff to maintain those servers.

Scaling up infrastructure on-premise can be cumbersome. Again, the need to configure the right hardware and software to support scalability can take up resources that you would rather invest in the business that the system supports. If you have been involved in planning, you would know that it takes lots of meeting sessions and numerous back-and-forth between the business and IT to decide the right system specification. With the cloud, you would still have these sessions between the IT and business but the decision-making is easier because you have the ability to configure your system on-the-go and on an as-needed basis. Do you need to scale up? You just need to have that budget ready.

Another scenario that shows the advantage of cloud vs on-premise is this: suppose that the business that the scaled-up system has changed and required you to scale down. With on-premise system, you would probably get stuck with the hardware that you no longer need and will continue to incur overhead because you still need to maintain it nonetheless. With the cloud, this is just matter of scaling down the service you are subscribed to. Scaling down is almost instantaneous.

If you have an on-premise system that needs additional compute power but are unwilling to spend additional dollars on hardware, you can utilize the cloud as extended compute node. You can take advantage of the flexibility of the cloud further by turning on or off the node according to your needs.

Mala: What are some of your favorite tools and techniques?

Marlon: On a day-to-day basis, I use SQL Prompt. My favorite feature on it, among others, is the ease of formatting T-SQL codes, snippets, and code analysis. I have saved all my frequent-executed queries as snippets and I could easily call these scripts on the fly when I need them. I can also configure SQL Prompt to format my codes according to our format convention.

SQL Monitor is another important tool that I use at work. If I get a ticket issue about deadlock, the SQL Monitor is the first thing I check. It provides essential info that I can use to research the issue. It gives you that capability to be truly proactive in managing your servers.

SolarWinds Database Performance Analyzer [DPA] is another tool that I regularly use. Combined with the stats you gather from SQL Monitor, DPA can further equip you with additional critical information to solve database problems and preempt database performance issues. DPA does not only monitor the overall performance of your servers but also the queries and query plans that are causing blocks or deadlocks.

Every time I get a report that a client's application is slow or a task in the daily job is taking longer than usual, aside from the monitoring tools, I would simply execute sp_whoisactive. This stored procedure can provide critical information about the current state of your database. To take advantage of the whole capabilities of sp_whoisactive, visit whoisactive.com.

All of our on-premise SQL Server deployments include the sp_Blitz suite. Checking the health of your server is as easy as executing sp_Blitz, sp_BlitzIndex, and sp_BlitzCache. A big part of my day-to-day task is ensuring that our SQL Servers are fast and reliable. The sp_Blitz suite gives you a good platform to manage servers proactively.

Glenn Berry's diagnostic scripts give you instant answers to questions about your SQL Server and the hardware that host it.

I always keep Paul Randal's Wait Stats query handy just in case I need it, and that is more often than I care to admit. If you are supporting applications, Paul's queries must be a part of your arsenal.

Lastly, you can never underestimate the power of the #sqlhelp hashtag on Twitter. The most fun part is when you post a question and a member of the MSSQL Tiger Team would answer.

Mala: What sort of struggles have you faced with managers/management, or the business side of things, and what is your recommended approach to handling that?

Marlon: The level of struggles that you would encounter between the business and tech people greatly depends on the depth of your organizational chart. The deeper the organization is, the more complicated the interaction is. Conversations can easily get lost in translation.

In any company with a traditional organizational chart in which there are multiple intermediary levels between the business and IT, changes are usually harder to adopt. The stakes are always high, and all the stakeholders need to sign off before things can decide upon. Decisions are made at multiple levels and at different phases.

Compare that to a non-traditional, flat organization where there is no clear demarcation between the IT and business. This is common for startups. Changes are easier to adopt and decisions are quickly made because your stakeholders and decision makers are often in the same group. Thereby, the channel in which conversation flows is shorter and so the decisions are made faster.

Obviously, handling different situations requires different approaches. If you are a part of a company with bigger, more complicated organization structure, the challenge is to get your message through the multiple levels quick enough so that the decision-makers get all the necessary information critical to making the decisions and clear enough so that the information does not get lost in translation. Provide the decision makers clear information about the what-if scenarios of every possible decision they can make. If possible, provide the information that could be reduced to require a Yes or No decision. This is difficult to do but not impossible.

In a flat organization, because decisions are easier and quicker to make, wrong decisions and mistakes are also easier to make. Prepare well your if-then scenarios and all possible outcomes must be provided to the decision makers at the same time the request has been made.

Mala: What sorts of issues are caused when you interact with other technologists? Which ones—developers, SAN admins, network folks—are you most likely to clash with, and why?

Marlon: Different roles in a good company are well defined and their tasks clearly outlined. Let the hardware people decide for the hardware decisions, and the software people the software decisions. The hardware and software decisions often overlap. This is where the Hardware people vs software people divide exists. There are decisions where what the hardware people think the software people need is not aligned with what the software people actually need.

This is a matter of getting the right information to the right people. You don't simply ask for "a disk" from the SAN admin when you need a physical disk with certain read/write latency spec and not a logical partition of an existing disk.

I believe that when you are clear about what you ask and why and provide the other party with the complete description of what you need, the other party will exactly give you that. Again, always be clear about what you need and provide information that will make their decision easier.

Another common issue that stems from interacting with other technologists is clash of principles. The traditional application developer may do things differently than the preference of the DBA. The importance of well-defined roles is critical in this situation. The default course of action should be to let the DBA decide for anything that will greatly impact the overall performance and health of the database. This is just an example but the same principle is applicable to the organization.

This is where an established company knowledge base and best practices come in handy. Most of these clashes among technologist can be avoided especially if similar cases have already been decided upon and recorded, for example, in a knowledge base or written in the best practices documents. We often heard about the phrase "single version in truth" in data warehousing. I think if we have a single version of the truth in all aspects of the business, clashes about who's right and who's wrong can be avoided.

Mala: What is the role of documentation in your job?

Marlon: So far, I have emphasized the importance of documentation and communication more than once in this Q&A. Like I said, one cannot just over-communicate. Documentation is a must. Not documenting is setting the next person up to fail. Not communicating well is a precursor to the disorganized team.

For example, if you are making some changes that would affect the overall system or application in one form or another, have those changes recorded in a document that is accessible to everyone that might need it. When documentation is not warranted, add some comments on the codes that would succinctly describe the changes. The next person will love you for this. Do documentation, either a formal documentation or comments, not only when possible but all the time.

Now that I am with an agile company, I finally understand the need for a version control system for databases. Changes that are made through the trunks and branches can also serve as documentation. Make it a requirement that comments are added in all commits.

Committing oneself to documentation and open communication is being responsible. And, I think requiring everyone to document is enforcing accountability. Documenting your actions is tantamount to saying that you have this ability to own up to mistakes.

Mala: What are your favorite books, blogs, and other means of learning?

Marlon: The SQLServerCentral blog is a site that I frequent. Having all the feeds from different blogs on various topics related to SQL Server makes SQLServerCentral a convenient platform.

I also love the blogs of Brent Ozar and his gang on their business/technical blog. Of course, I also love reading Brent's insights on business and life on his personal blog, ozar.me.

If I need knowledge about SQL Server internals, I would go to the SQLskills blog by Paul Randal et al.

For books, I always make sure I have the latest version of Itzik Ben-Gan's *T-SQL Fundamentals* [Microsoft Press, 2016]. Dmitri Korotkevitch's *Pro SQL Server Internals* [Apress, 2016] can be overwhelming at times, but if I'm in the mood for seeking knowledge about SQL Server internals, I open a chapter or two. My recent SQL Server book favorite is the one written by Benjamin Nevarez, *High-Performance SQL Server: The Go Faster Book* [Apress, 2016]. I am learning a lot about service-level requirements, datastore architecture, release management, and more from *Database Reliability Engineering: Designing and Operating Resilient Database Systems* [O'Reilly Media, 2017] by Laine Campbell and Charity Majors.

For non-technical books, I love Ray Dalio's *Principles* (Simon & Schuster, 2017), which outlines his approach to decision-making in business and in life, and his philosophy of "idea meritocracy"—strive for meaningful work and meaningful relationships through radical truth and radical transparency. Another non-tech book that I love is the recent book by Tim Ferriss, *Tribe of Mentors: Short Life Advice from the Best in the World* (Houghton Mifflin Harcourt, 2017), which has the same format as this book—that is, basically in Q&A format.

For further learning, I would attend conferences, specifically SQLSaturday events. For online learning, I have a subscription with Pluralsight and Udemy. Another site that I like is Channel 9 at the MSDN site.

Mala: What are your recommended ways of stress management and developing healthy work/life balance? How do you handle long grinds on the job?

Marlon: In the tech universe, work/life balance is a unicorn. You may be on a scheduled vacation with your family but an emergency at work would suddenly require you to call in, or worse, to log in at work to solve issues. Or, you may have to excuse yourself from work if and when your kids need you at school. This is the normal life of the tech person.

The challenge is not to achieve that elusive perfect work/life balance, because that is quite impossible to have, but to achieve a certain level of balance where you are capable of fulfilling your different sets of responsibilities with your work and life without compromising both at the cost of one over the other.

If you are juggling different priorities that are of equal importance to you, managing your time is quite impossible. We only have twenty-four hours in a day. Normally, one-third of that is spent on work. But if you are in the tech industry, you probably need more than one-third to effectively fulfill your responsibilities.

Instead of managing time, manage your priorities. Family always comes first, as they say, but that doesn't mean you have to choose between work and family. This is why knowing your priorities at any given time is important.

Make separate lists of your work and personal tasks according to priorities. This way you can have an overview of how your general priorities look like. At times, you cannot avoid having a conflict between personal priorities and work priorities. You should be willing to make tradeoffs and be ready to deal with the outcome of those tradeoffs.

At work, I use our ticketing system to manage my work. I usually augment that with the my to-do list. An old-school pen-and-paper list. Usually, a sticky note pasted next to my laptop touchpad.

We are always grinding in our tech jobs, which is why health should also be one of our top priorities. I take supplements and vitamins. Eating healthy and doing physical exercises are two areas that I need to improve on. These are essentials when it comes to endurance and overall health.

Some specific part of a tech job can sometimes become monotonous, especially when you're doing the same thing over and over again. This can lead to a state we call "being stuck in a rut." This can be avoided by presenting yourself with new challenges. Ask for a different assignment or take courses to improve your skillsets.

A good outlet when boredom strikes is getting yourself involved in the non-tech community, like charity institution, or community service.

Mala: Describe your style of interviewing a data professional. What do you look for, and what are some examples of questions you ask?

Marlon: I focus on basics and fundamentals, like indexes, execution plans, locks, and deadlocks. I want to make sure that they know the basics.

Another big thing on my list would be the concepts or principles in database management or development, like how to design the database or how to write the stored procedures to avoid possible unnecessary locks and deadlock or how to architect a system for performance and scalability. I know these are broad topics, but I want to understand the candidates' philosophies on these topics.

I would also like to know how they manage their time and tasks. I want to know if they have skills to manage their job and the customers that we are serving. Efficient employees are good managers.

I really do not mind if they cannot memorize management scripts, as long as they know the right data management views or procedures to use depending on the situations.

I also want to know if they are team players and, at the same time, can work independently. This is critical especially when the job requires the candidates to work with teams.

Another question that I would ask if they are willing to work extra hours and as needed on weekends or holidays. This is a reality that we face on tech so I just want to make sure that I am communicating this to the candidates on the onset.

I want to know the personal initiatives the candidates are taking towards learning. Having the personal initiative and willingness to learn is one of the most important characteristics that I want to see in candidates. I would also like to know if they are aware of other technologies other than the ones they are currently using. That way I may know I have people on my team that can serve as good resources if and when I need information about technology options that we want to explore.

Mala: What are your contributions to the community, and why do you recommend people be involved with the community?

Marlon: I blog when my schedule permits. This is one venue that I think I am contributing to the tech community. I also wrote a book which I think is another contribution to the community.

I can also contribute to the community by adding values to meaningful conversation on social media about important topics in technology.

Speaking at an event could have the most impact among other venues. If you have the speaking skills and can deliver a session or two to a tech event, you should take that opportunity.

I am a benefactor of the great contributions of lots of SQL Server professionals. In fact, it would not be an exaggeration if I say I owe my career to the tech community. Being a part of the community does not only give you that sense of belongingness but also empowers you to want to improve yourself and profession.

I found inspirations in people who have improved their life and career and elevated their careers to unimaginable heights. I could not have witnessed these if I weren't part of the tech community.

Getting involved in the tech community can only do you good. Reaping the benefits of being part of the community is almost sure. Plus, you wouldn't who you inspire or help. The best part is every time I receive a message thanking me for the blog post I have written because it helped them in one way or another.

Mala: If you had one superpower what would it be, and why?

Marlon: Being able to turn back time. I do not regret the things that I have done in my life but if I can turn back time, I have things that I would like to do differently.

Key Takeaways

- The most important thing in career planning is knowing exactly what you want to achieve. So, the adage "failing to plan is planning to fail" (often attributed to Benjamin Franklin) is not only true to a sense but also true when it comes to one's career.

- Knowing the principles of database administration is more important than knowing how to manage the database in either Oracle or SQL Server. Equip yourself with principles and techniques instead of the nuances specific to technologies.

- If you work in any data profession, you are not an island, as no man is. You cannot operate as a lone professional in your own silo. You are part of a team, even when you are the lone company DBA or SQL developer.

- How far you advance in your career greatly depends on how much effort you are willing to put into learning.

- If we have a single version of the truth in all aspects of the business, clashes about who's right and who's wrong can be avoided.

Favorite tools: Redgate SQL Prompt and SQL Monitor, SolarWinds Database Performance Analyzer (DPA), Adam Machanic's sp_whoisactive, sp_Blitz, Glenn Berry's diagnostic scripts, Paul Randal's Wait Statistics

Recommended books: *T-SQL Fundamentals* by Itzik Ben-Gan, *Pro SQL Server Internals* by Dmitri Korotkevitch, *High-Performance SQL Server: The Go Faster Book* by Benjamin Nevarez, *Database Reliability Engineering: Designing and Operating Resilient Database Systems* by Laine Campbell and Charity Majors, *Principles* by Ray Dalio, *Tribe of Mentors: Short Life Advice from the Best in the World* by Tim Ferriss

Blogs to follow: SQLServerCentral.com, SQLskills.com, Brent Ozar's blog

Joseph Fleming

Senior Consultant, StraightPath Solutions

From the moment his fingers hit the keyboard on the TI 99/4A that his dad brought home, **Joseph Fleming** *knew computers were destined to be a big part of his life. From typing hex-based programs out of Compute!'s Gazette magazine on a Commodore 64, to his first programs in BASIC, it was just a matter of where destiny would take him.*

A brief fling with the idea of becoming an automotive engineer almost waylaid him, but a career in IT was meant to be. It began with a job working on a proprietary database system (through a connection at a kung fu studio—you never know where opportunity is going to knock) and learning how relational database management systems worked. He started working with SQL Server in 1998 using version 6.5; he has battle scars from upgrade cycles that may or may not have followed best practice guidelines. His skills lie in performance tuning, upgrades, SQL Server replication, and high availability.

Joe has strong ties to the SQL Server community through social media and PASS, serving as a PASS chapter leader, regional mentor, and SQLSaturday volunteer and organizer. He has also mentored several groups of PASS Summit "first timers" looking to make the most of their experience. He posts on Twitter as @muad_dba and his blog is at muaddba.wordpress.com.

© Malathi Mahadevan 2018
M. Mahadevan, *Data Professionals at Work*,
https://doi.org/10.1007/978-1-4842-3967-4_27

When he's not cursing at code that he's trying to debug or optimize, Joe enjoys wallyball, outdoor adventure parks, skydiving, and coaching and volunteering with his kids.

Mala Mahadevan: Describe your journey into the data profession.

Joseph Fleming: My journey into the realm of data started the way I think a lot of them do—by accident. I was pursuing an MIS [Management Information Systems] degree and looking for a computer-related job—any computer-related job. I could have very well ended up in computer repair or as a programmer, but as luck would have it, there was a medium-sized company in my area looking for some entry-level help with data processing. At the time, it was called The MEDSTAT Group—it's now part of IBM Watson Health—and the goal of my department was to help companies manage their rising healthcare costs by analyzing their claims and enrollment data. This was my first real exposure to databases as opposed to a simple spreadsheet. It was an internship that converted into full-time work at the end of the summer, and it's still a place I look fondly on in the rear-view mirror. I met my wife there, as well as one of the groomsmen for my wedding, and many other really talented people.

My first exposure to SQL Server databases came from my next employer, where I did telephone support for another healthcare-related product that was multiplatform—SQL and Oracle. I learned a lot about both, but I found SQL Server much easier to understand and configure, and soon I was traveling to customers and helping them install and configure our software.

All of that experience helped me to land my first full-time DBA job in the year 2000, just as the super-hot employment market was drying up. I was happy to land at a place that had a reputation for never having rounds of layoffs at a time when rounds of layoffs were becoming more common.

I was so excited to finally be called a database administrator! I was the only one on the team, and stepping in for someone who had already left, so there wasn't much in the way of transitioning, but I dug right in and started fixing things, like index maintenance and other daily tasks that were a little out of whack, just based on common sense and what I could find in books online, because the community back then was nothing like it is today, with instant help available via Twitter, Stack Exchange, and various other forums.

It was at this job that I learned to dread the is-everything-okay-with-the-database question. It's so common and always leaves me with this sick feeling. The problem is rarely the database, but it's always the first thing anyone asks about. And back then, I always felt like I had to run through the full laundry list of possible issues before I could definitively say, "No. It isn't the database." We had our share of performance problems. One day, my boss took us all over to the data center with the order to stare at the servers until they start behaving. And, yes, I am serious. We did that.

Mala: That is hilarious! And?

Joe: Yep. In all fairness to him, he was a really smart guy who helped me solve many database problems—although he wasn't a database guy—just by talking through them and asking great questions.

Alas, all good things come to an end, and my stable company decided to join the ranks of other companies in outsourcing and offshoring. I was given an end date, about nine months away, and told that I should find somewhere else to work. It was a blessing in disguise because leaving there in 2006 helped me discover the wonderful SQL Server community at my next job. My boss encouraged me to get involved in the local user group and sent me to my first PASS Summit in 2007. I worked on a team of twelve DBAs, which helped me learn a lot faster than before, and soon I was on the path to becoming a chapter leader, and eventually a regional mentor, helping to spread SQL knowledge and broaden the SQL community.

I've had a few more jobs since then, but these are the ones that really established my career and got me where I am today.

Mala: Describe a few things you wish you knew when you started your career that you know now and would recommend newcomers to this line of work know.

Joe: Looking back, there are a lot of things I wish I had known when I started off as a data professional. I wish I had known better early on how to be both assertive as well as empathic. Often times, I was either a pushover or a sort of arrogant stand-your-ground type, and I needed a lot of time to figure out how to project confidence without appearing too idealistic. The truth is that everyone's just trying to get a job done, and you have to deal with them as people and not "evil developers" or "unrealistic users."

It might seem surprising, but I wish I had realized just how much more powerful set-based operations are earlier on. During my first few years as a DBA, I was still stuck in the very procedural way of thinking about things that you learn when you're at school. At least back when I was in school, nobody really taught you about set-based operations for databases, so I can definitely see why it's one of the things that cause friction between software developers and seasoned DBAs.

Another thing I wish I had learned earlier—and I still struggle with now—is that while electronic communication can be wonderful, linking people who are thousands of miles apart, it can also be dehumanizing. You come to see the person on the other end not as a person but as just another electronic reminder to respond to. You have to pick up the phone or walk down to an office and have a person-to-person conversation to remind yourself that these other folks are humans. And sometimes they need a reminder that you're human as well and not just a robot that they feed tasks to. For all the flexibility that comes from working remotely, there is a cost. It's in the extra effort required to make sure you're not becoming too reclusive, reactionary, and insensitive to remain effective.

I wish I had discovered bloggers and social media sooner. After my second PASS Summit, I got on Twitter and realized just how much help was available and how much I could learn just by watching the feed of other people asking for help. I was quickly connected with a lot of really smart people who genuinely wanted to help me solve my problems and learn. It wasn't just learning about SQL Server stuff, either. One of my favorite presentations to this day, which came at a time when I really needed to hear it, is a five-minute talk by Buck Woody called "Your Career Is Your Fault." He turned it into a blog post that you can find pretty easily with an Internet search. The bottom line is that you are in charge of your career, and if you let other people drive it, then you're going to end up going where they want you to go. And they probably won't let you change the radio station either.

Also, don't shrink the database. Or use cursors.

Mala: Describe how a DBA manages to maintain his/her policing/gatekeeping duties with frequent deployments that come with an agile environment?

Joe: For a DBA, dealing with the agile development methodology can be a real nightmare. Part of that is because it's often misused by folks who think it means that you can spend minimal time in design and just work it out as you go along. The DBA can get overwhelmed either making incremental changes, where the cliff of doom ahead is obvious to all but those rushing toward it, or caught up constantly saying no and being overridden. Another part is that it goes against the way most DBAs tend to think. Remember, we like things done in a set-based approach. You take a large chunk of code, and you test it all and put it into production. Constantly making small incremental changes goes against the grain of how DBAs are trained to think.

I think the secret sauce lies in helping to educate the development team and being open to being educated yourself. Developers have to understand the principles of good database design and what happens when you don't have it. Having a published list of standards will help. I worked at one employer who had a forty-page document of SQL Server design and coding standards. Some people gasp when I tell them that, or insist that it's ridiculous. But those folks aren't responsible for supporting a sub-second online cloud service either. And DBAs have to understand the impact that a properly implemented agile process can make when it comes to the finished product. The users are happier because they have been able to provide feedback throughout the process instead of just at the end when the finished product is delivered "according to the specifications."

Another ingredient in the secret sauce is tailoring the response to the situation. A forty-page standards document would be ridiculous in many environments, but it's absolutely imperative in others. Know your environment and know the important battles, because fighting wars on too many fronts simply leads to having your enemies use flanking maneuvers. Did I say enemies? I meant co-workers. Often, showing people why a badly designed process won't work out in the long term will be enough, but there are always folks who have deadlines and bonuses to think about and will insist on forging ahead. I've seen this result in serious damage to the reputation of the IT department, to the point that it caused management shake-ups. It also means no end of headaches for the DBAs and others in support roles. You have to think strategically. Complaining that standards are not being followed won't be good enough. Show what kind of damage those violations can do, or has done, and how simple it is to correct.

In the end, I think that considering yourself to be a "gatekeeper" can be part of the problem. You're part of a team tasked with delivering a quality product to a customer. Working as a team will get you a lot farther than positioning yourself as some sort of enforcer.

You *can* change some of the frustrations of working in an agile environment, but expect it to be a long process. A well-run agile process will still be fast-paced and leave less time for review than most DBAs would like, but with everyone listening and learning, the output will be something everyone can be proud of.

Mala: Describe some of your favorite tools and techniques.

Joe: As a DBA, I use a lot of tools in my day-to-day activities. In no particular order, I'll list some of my favorites.

Google! As crazy as it might seem, knowing how to use a search engine well will make your job a whole heck of a lot easier. It helps to have a big vocabulary and to be as specific as possible. Consider the search engine like you would consider someone learning your language. Be prepared to offer the search engine several different phrases that all mean the same thing, and then learn which sites are most likely to return helpful results and which ones are more likely to just be noise.

I also like to use some of the simple built-in T-SQL troubleshooting tools. sp_who2active will give you a quick look at the active sessions on the server without the overhead that Activity Monitor will generate. When you find a SPID that you think might be problematic, DBCC INPUTBUFFER will give you a shortened look at the executing statement. DBCC OPENTRAN is one I use a lot to find out what's using up all the space in my tempdb. Knowing these built-in functions is critical when you're working on many different customers' systems, and they don't all have some of the neater toys installed.

Another great free tool when you're managing a lot of different servers is Remote Desktop Connection Manager. Like the Registered Servers window in SQL Management Studio, it allows you to create groups and subgroups of servers that you typically use Remote Desktop to interact with. You can quickly swap between them without having to use Alt-Tab or find the right one on your taskbar. And you can save your configuration to a file and quickly deploy it to other folks on your team, or to new team members when they start.

I'd be surprised if this next one didn't make the favorites list for several people. Adam Machanic's sp_whoisactive. A tremendous improvement to sp_who and sp_who2, it can be configured to give you a great deal of information, including something I find really handy—the query plan for each running query. If you don't already have it installed, read up on the things it can do and see about deploying it in your environment.

I think my favorite technique is using SQL to create more SQL. If you're not familiar with that, I'll give a simple example. Let's say you're in a development environment and you want to clear out all your tables and reload them. You could type out a bunch of TRUNCATE statements, or you could use what you know about system tables to make it a little easier on yourself. If you concatenate some text [TRUNCATE TABLE] with the NAME column from sys.tables, you can quickly build a truncate statement for all of the tables in your database, or limit them with a WHERE clause if needed. Once you've mastered some of the simpler ones, you can start getting creative, using CASE statements to tailor commands to different situations. It has come in very handy in a lot of situations, saving hours of typing.

Mala: As a DBA, what issues are caused when you interact with other technologists? Which ones—developers, SAN admins, network folks—are you most likely to clash with, and why?

Joe: Different people, different technologies, and everyone believes they know the truth. I'd say that's the biggest challenge that comes from dealing with other departments. Let's try to break it down into some of the different groups.

I think developers are the most challenging because there are usually a lot of them. Whenever you get a large group, it will be harder to manage. It's harder to manage expectations, and it's generally harder to get them to coordinate with each other. A typical company will have development teams split into different products or lines of business, and there is not a lot of coordination that occurs until the change management meeting the week before they deploy. Then comes the chaos as everyone scrambles to make sure that other projects don't impact them, and there is shuffling of priorities and fighting over which project is more important.

And it's going to be really strange for some people to hear this, but I've actually had the pleasure of working with really great storage people no matter where I went. Typically, they were willing to listen to my opinions on managing storage for databases and were very helpful and responsive when it came to things like allocating new space and planning for disaster recovery.

Mala: Wow!! You are lucky there. What about network engineers?

Joe: I respect the heck out of the people who work on networks, because if there is one area that probably gets asked if there is something wrong with the technology more than DBAs, it's the network guys. And it's probably even less likely to be the network than it is the database. And it's just as frustrating to hear the automatic "it's not the network" for a DBA, as I imagine it is for the people who have to hear "it's not the database."

The best way to interact with other infrastructure people is to learn a little bit about their technology and how your technology uses it, so you can ask intelligent questions. Instead of asking if there is a problem with the network, you might ask if there is any packet loss happening between server A and B. That way, you are asking them some specific questions instead of sending them on a wild goose chase. Once you have exhausted your knowledge, or when they tell you to stop asking dumb questions, you can ask, "Can you think of any other reason for problem X that could be network related?" Maybe they will have a helpful suggestion that doesn't involve a place where the sun doesn't shine.

In the end, all of these folks are people, and your challenges are more likely to revolve around what kind of people they are than which technology they deal with. There are folks out there who refuse to listen, refuse to learn, generally make your life miserable, and somehow manage to keep their job for several years. They make you want to belt out a string of expletives or break things— or both. Hopefully, your interaction with these types of folks will be brief.

Treat people like, well, *people,* and they will generally respond in kind. Help them to understand that you're all on a team, and that you're all working toward a common goal, and the good ones will understand and collaborate. Learn to recognize that there are also folks who will take this teamwork approach to try to get you to do their job for them, or to bypass important standards "for the good of the team/company." Don't let them push you around, because that will only make it easier for them to do it the next time, and harder for you to say no.

Mala: What has been your experience as a DBA and a consultant with cloud adoption?

Joe: The cloud! It's coming! I've been hearing this refrain for years, and it hasn't really become the career-shattering event that the hype proclaimed it would be, at least not for DBAs. Having said that, it really is changing the way business is done across many sectors.

While a DBA, I watched some development teams move their dev environment into Azure because it's pretty cheap, and then they can manage their own servers there, which is not necessarily a good thing. While there are some downsides to letting developers manage their own infrastructure, there are some really huge upsides. Because they manage it, they have to support it, and I generally didn't get called overnight to help someone fix a broken Azure dev environment. When they have to support it, they learn to build more robust systems, and that's an advantage for everyone involved.

I also worked for a couple of companies that essentially provided cloud solutions, like product lifecycle management and ERP [enterprise resource planning]. Some of them were built on cloud platforms like AWS, and others were in-house-designed cloud platforms that had tremendous performance expectations.

As a consultant, I've helped companies leverage some of the flexibility that the cloud can offer. One project I worked on helped a company to know when it was time to add more capacity to their cloud setup, or when to migrate servers with demanding workloads away from shared resources.

The cloud can offer tremendous flexibility and cost savings, but it can also be a black hole that sucks your projects in, never to be seen again. If your company is thinking about cloud adoption, they will be tempted to simply spin up some resources and get to work, because it's super easy to do that. Don't.

Things in the cloud need to be planned out well for the same reasons that things in an on-premise infrastructure do. Without good planning, you end up committing to certain directions that can be incredibly difficult to change after a year or two—sometimes even just a couple of months—of continued development.

There's a saying I learned at one of my prior companies. "We never think there's time to do it right, but we always think there will be time to do it over."

The v2 release will never have the budget or the time built into it to fix all the problems introduced if v1 was not planned and implemented properly. It's true whether you're doing development or infrastructure planning. Do your best to get it as right as possible the first time, and build on a solid foundation. If you don't, you will end up using all sorts of shortcuts and hacks to keep things running. Those shortcuts and hacks will eventually collapse under their own weight, and everyone will wonder why you didn't just build it right the first time.

Mala: What are some industry trends in this line of work that you are particularly excited about and watching evolve?

Joe: As a DBA, one of the most exciting trends I see within the industry is that people are becoming more aware of their data. I think one of the biggest benefits that have arisen because of social platforms like Facebook and technology like smartphones are that people are starting to think more about how their personal data is being used, and they want the folks using it to be more accountable for it.

Data security is a *huge* problem. Rather, I guess I should say that the lack of data security is a huge problem. It seems like new data breaches are announced monthly if not more often, and from places like Experian that hold critical information about so many people.

More and more tools are becoming available to help data privacy stewards manage PII. PII is *personally identifiable information,* which is data that identifies you. While sometimes related to hijacking credit cards or bank accounts, PII can also be used to harass or intimidate people. Two examples of this are doxing – publishing a person's location/contact information on the internet so that harassment comes from hundreds or thousands of folks, and swatting – calling the police department and reporting a highly dangerous situation at someone's location so that the SWAT team shows up.

Yet still, a majority of companies simply make copies of their production data to use for dev and QA unencrypted and unscrambled, not thinking about the risk involved with giving tens or hundreds more people access to that data.

It's good to see companies being held accountable for lax practices when it comes to securing the data of the customers who place their trust in them, and I would love to see some sort of market-driven certifications to show that companies are not being stupid with customer's data.

There are dozens of companies now that are involved with data masking and encryption, and SQL Server has some great built-in features to support this as well. It's a huge and daunting proposition to try to bring data protection precepts into a company without them. There are so many challenges, from its impact on the speed of development, to the additional red tape involved when getting access to unmasked data, but in the end, I think it is well worth the additional costs if we all end up with a better understanding of how to protect ourselves.

Mala: What is the importance you place on documentation for DBAs, and how do you approach it?

Joe: I think documentation is a tremendous challenge for professionals in every walk of life. It can be tedious, boring work—not nearly as exciting as solving a neat new problem or attending a seminar, if you're into that sort of thing. The reward for a job well done is often that people ignore the document or it is placed somewhere that's hard to find. It can be really disheartening. But, it plays an incredibly important role in how successful your department and your company will be. Good documentation helps to transform the little knowledge that someone might have about a process, a customer, or a piece of technology, and turn it into understanding.

Mala: It has to do with good writing too. Don't you think? I think writing well is important. Documentation can be interesting or more readable if written well.

Joe: Yes, absolutely. I am sad to say that writing—especially good writing—is just not valued by many folks anymore. You don't have to work at a company to see it. You can just look at your local newspaper. Many articles are poorly written and edited because people won't pay for newspapers anymore, so writing and editing staff are cut to the bone. Within companies, writing staff are often ignored or not given the time they need to produce good documentation, whether that time is their own or time they are supposed to have with the technical folks who are supposed to help them put to words the brilliant new solution they have designed. And that assumes that the company actually has a writing staff. Often it is simply part of the duties assigned to technical resources, whether they can do it well or not.

Good documentation isn't just about being able to follow the rules of grammar. It's also about being able to read what you've written as if it didn't just come out of your own head. You have to be able to look at it and think, "Is there any part of this that could be read in a way different from my intent?" If the answer is yes, try to make it clearer without turning a single paragraph into a twenty-page dissertation. You don't need to make it accessible to an elementary school student, but if your writing looks like an alphabet soup of acronyms and abbreviations, it might be time to take a second look at it.

It's also about being organized. Just putting all of your thoughts onto a page does no good if there isn't a logical flow to them. It's interesting that so many folks who can write amazing code, which requires planning, organization, and flow, seem to fall apart when asked to do the same thing with documentation.

Okay, so you've got a great document. It's grammatically correct, it flows like one of Shakespeare's sonnets, and is chock full of useful information. The last part of the good documentation is in making it easy to be found by those searching for it. If it's buried under twenty-five layers of folders in an incomprehensible structure and given a name like "Document 1," then it's like trying to find Amelia Earhart's plane. It's going to take forever, and by the time you find it, the world will have moved on.

Mala: What are your recommended ways of stress management and developing a healthy work/life balance?

Joe: One of my favorite comics is *Calvin and Hobbes*. In one comic, Calvin's dad is preparing to attempt some plumbing repairs. Reading from his do-it-yourself book, he says, "Before beginning any home plumbing repair, make sure you possess the proper tools for the job. Check the following list of handy expletives, and see that you know how to use them."

Sometimes you really just want to curse at the top of your lungs, which is difficult in today's world of cubicles. I had one co-worker who invented fun new phrases to express his frustration, so occasionally I'd hear him yell, "Monkey balls!" from over the cubicle wall. As a consultant, I have to constantly remind myself that if the customers' problems were easy fixes or their environments simple to manage, they wouldn't need me, and I'd be out of a job.

Having a release valve or two for that stress is essential to good mental health. I find that games that let you release pent-up aggression are great for that. I play wallyball, which is like volleyball but played on a racquetball court. It's fast-paced, and I get to hit the ball really hard and get too exhausted to dwell on the frustrations at work. Afterward, we go out for drinks, so I get double the stress relief.

Running is also good because I just sort of zone out while running and let myself be caught up in the exhaustion and the pace, and how much distance is left until I can stop bloody running.

Another great way to reduce stress is to find your local user group meeting and just talk with other professionals who may be suffering through the same things you are dealing with. They might even have some helpful advice.

One thing that I have found is definitely not helpful is to sit around with your co-workers and gripe about all of the bad situations. The negativity feeds on itself, and before you know it, everyone's even grumpier than before. As corny as it sounds, the time is better spent trying to figure out ways to make things better. I mean, there is some merit in trying to figure out if maybe you're just in the wrong place, but you have to realize when you have moved beyond venting and into something demoralizing.

Attaining work/life balance took a while for me and can still be a challenge from time to time. No matter what job you're at, there's almost always going to be someone who is willing to work longer hours and give up more time than you. Or maybe they are one of the rare people whose job is also their passion and hobby. Even if you have no major plans for the night, it can be intimidating to have to tell that person, "Sorry. I have to head home."

Sometimes you'll work somewhere that people regularly work sixty-hour weeks. You'll feel like a slacker if you don't do the same. You're not a slacker. The company simply has too much work to be done given the number of resources they have available. If everyone just keeps sucking it up to meet the deadlines, nothing will change. That guilt doesn't belong to you. It belongs to the folks doing the scheduling and setting the deadlines. Don't try to own it. Instead, you have to be secure in the knowledge that you're doing your best. After all, for most of us, *work* is what we do so we can do other things, and that shouldn't fill you with guilt, but it still will.

Mala: Describe your style of interviewing a data professional. What do you look for, and what are some examples of questions you ask?

Joe: Interviewing is hard! I've been on both sides of the table quite a bit, and I think that giving an interview is just as nerve-wracking as being interviewed. Actually, having the right answers to your technical questions is a very small part of whether or not a candidate will be successful. Nevertheless, ask those technical questions, because they are going to start the conversation you really want to have. The one that begins with *why*.

When I am interviewing someone, I am really looking for a few key things that will indicate their ability to succeed. First, does this person have the ability to logically reason through a problem, whether it's a simple logic puzzle or a complicated database problem? At a basic level, can they use a search engine effectively? I've worked with people who can't, and it's an important skill when the platform is as broad as SQL Server is. Can they think, or do they just know the answers they read in a book somewhere? Memorizing isn't all bad. A lot of the problems I have solved have been because I remembered something I learned either from a book or a forum post, or something else, but when it's three a.m. during an all-hands-on-deck scenario, and you've encountered a problem that Google doesn't have the answer to, you need to be able to try to think up new ideas for what could be wrong and how to solve it.

Second, can this person admit when they don't know the answer, instead of feeding me a line of bull? It's perfectly okay to not know the answer to every technical question you're asked. What's not okay is to pretend that you know the answer and stubbornly try to make your answer fit the question by weaseling. Often, I will try to push someone into a situation I know they don't know the answer to, to see how they react. Some folks don't like this and think I am trying to make myself feel superior. I got a nasty e-mail about that once. But when push comes to shove, when the database is down, nobody really cares that you were right on a technicality. They care that you were able to answer these questions: How do I get it back up again? And, how do I prevent this from happening next time?

Finally, can they admit they have made a mistake? I always ask interview candidates what the worst thing they have done to a production system is. If they refuse to answer the question, I will admit to some of my own failures to help them see I am not infallible, and I don't expect it from them. If they insist that they have never made a critical mistake in production, I'll generally pass on that candidate, because they are either lying or they've never been in one of those gut-wrenching situations where you caused the problem and now you have to solve it and take the heat for it. It would be like hiring a boxer who's never been punched in the face. I'd always wonder how they would react.

In the end, your success as a DBA is going to rely on knowledge, relationships, and integrity. If you can fake all three of those things during an interview, you'll probably get hired. Then you just have to fake them for the rest of your career.

Mala: What are your contributions to the community, and why do you recommend people be involved with the community?

Joe: I've been involved in the SQL Server community for about twelve years now. Ever since attending my first user group meeting, I was hooked on the idea. I was fortunate enough at that time to have a boss who encouraged me to become more involved in the user group and who sent me to the PASS Summit. I tried to become involved in the local group, but I was having a hard time. I wanted our group to grow and reach out to more people, and I knew that we needed serious effort in terms of courting sponsors and speakers in order for that to happen. Microsoft was stepping back from single-handedly funding our local meetings, and we needed to seek new ways to make them happen. In the end, I chartered a new group in the area and partnered up with one of the local Microsoft TSPs [technology solutions professionals], Rick Brewis. I was so fortunate to have him helping me navigate my way through the start of the group. We kicked off the group with a December holiday party, and things grew pretty quickly from there.

We grew from about twenty-five regulars to around forty-five per meeting. My boss did a remote meeting. I was introduced to Jeff Moden, a local Microsoft SQL Server MVP who was a fantastic presenter. It was great to bring more and more knowledge and mentorship to local people. I got more deeply involved in the PASS organization, and became a regional mentor, helping other folks who were forming groups or trying to grow existing ones in the Michigan, Indiana, and Ohio areas.

It would never have been possible without Rick's help, so I will always be grateful to him. Another person I will always be grateful for is Karla Landrum. As user group growth spread, sponsorship money was not quite what it once was. She gave me some great ideas for saving money and getting the user group running on a tighter budget. And when I decided to try to launch a SQLSaturday in my area, she was constantly available for questions and ideas for running it.

But the community is more than just PASS, user groups, and SQLSaturday. It's the people involved who volunteer their time to help educate other people with no reward attached to it. I've been involved with a lot of different technologies, but I have yet to find one as supportive and familial as the SQL

Server community. At one point in my career, I was offered the opportunity to move from SQL Server and focus on PostgreSQL instead. A major part of my decision to decline was just how awesome the community surrounding SQL Server is and how welcoming the people within it can be. Sure, there are bad eggs in every community, but if you look, you'll find a place to fit in here, and the rewards will be well worth the time you spent.

You'll make new friends who span the globe, and you'll chat with them mostly via Twitter or e-mail. If you're lucky, you'll get to meet them at a conference and tell them just how thankful you are to have them around. I'm fortunate enough to have been able to thank many of these folks in person, and despite the "celebrity" status of many of them, they have all been humble, happy to have helped, and responded with a smile and a handshake.

Mala: Yes. Some of us are pretty fortunate that way. What would you have to say to people who are not naturally extroverted or outgoing?

Joe: I've been told that I'm "pretty outgoing for a DBA" on more than one occasion. It's not easy for me to be outgoing. I default to an introverted person who would much rather sit at home and watch TV. You need to push your comfort zone. It's only because of other great, outgoing database folks that I've been able to make as many friends in the community as I have. If I can do it, you can probably do it. The community is there, just waiting for you to join it. I don't think you'll regret giving it a try.

Mala: Can you narrate a funny/interesting story to share with our readers? It does not have to be work related.

Joe: Sure. At my first DBA job, I had a boss named Steve. One of his first tasks for me was to have me build a report telling him which jobs ran or were running during specific times of the day for each of our servers. I spent several hours working on the report, looked it over, and then saved it and e-mailed it to him as an attachment. I called the attachment "Steve's Report".

As our team grew, Steve would ask each of the new people to create some sort of report for him. One day, in a meeting, he asked our newest team member to build him some sort of report based on their area of expertise. He added, "And, for heaven's sake, don't attach it to an e-mail and call it 'Steve's Report,' or I'll have to fire you. I hired all these really smart people, and every damn one of you names your first report after me and sends it via e-mail. Give it a meaningful name and put it out on our file server for crying out loud!"

Everyone—except the newest recruit—looked at each other and laughed knowing that we had all made the same mistake.

Much later, I made a mistake that ended up costing the company about double my yearly salary. It was a substantial amount of money for this company, and the bonuses that were typically paid out each quarter were significantly lower because of my mistake. To top it off, it was right around the holiday season, and I am sure many people were counting on that bonus to make holiday expenses easier. How I reacted to the mistake is what kept me working for them for another three years.

I accidentally ran a script in production that was supposed to be run to clear out the development environment for a new test. It was disastrous. As I walked down to the office of the senior person on the team, I heard him in the director's office, saying he wasn't sure what the problem was or why things weren't working. I walked in and told them it was me, and owned up to what I did, and we worked together—along with a lot of dedicated developers, storage admins, network admins, and web admins—to get things back up and running.

What I said about integrity earlier, it's important. Most companies I've worked for wouldn't fire you for making a mistake, but they would fire you for trying to cover it up.

Okay, a funnier one. This one could have been from a movie. It was winter in Michigan, so the traffic on my morning commute was slower than normal. I was late for a meeting, so I dialed in from the car. I was driving in the right lane so I could go slower and respond to one of the questions during the meeting. A salt truck had just entered the freeway ahead of me, and it began dumping salt out the back. The salt was bouncing off the road and hit my car and my windows, so I slowed down. The guy behind me flashed his light and honked his horn. I screamed at the car. It was a very colorful description of something unpleasant I felt the other driver should do. It was suddenly very quiet on the other end of the phone. Very quiet. This was post Y2K when the job market had dried up quite a bit and finding a new place to work was not going to be easy. I apologized and explained what happened. There was some chuckling on the other end. I breathed a sigh of relief. Then, one of the folks in the conference said, "I've never heard that particular phrase before. It was really awesome. I am going to bring you a cheesecake!" And that's how I got a cheesecake for swearing during a conference call.

Key Takeaways

- Everyone's just trying to get a job done. You have to deal with them as people and not "evil developers" or "unrealistic users."

- You are in charge of your career, and if you let other people drive it, then you're going to end up going where they want you to go. And they probably won't let you change the radio station either.

- Working as a team will get you a lot farther than positioning yourself as some sort of enforcer.

- Good documentation isn't just about being able to follow the rules of grammar. It's also about being able to read what you've written as if it didn't just come out of your own head.

Favorite tools: Adam Machanic's sp_whoisactive, Remote Desktop Connection Manager, using SQL to write more SQL

I

Index

Printed in the United States
By Bookmasters